ENERGY POLICIES, POLITICS AND PRICES

JUSTIFICATION OF LONG-TERM ECONOMIC POLICY OF RENEWABLE ENERGY SOURCES

ENERGY POLICIES, POLITICS AND PRICES

Additional books in this series can be found on Nova's website under the Series tab.

Additional e-books in this series can be found on Nova's website under the e-book tab.

ENERGY POLICIES, POLITICS AND PRICES

JUSTIFICATION OF LONG-TERM ECONOMIC POLICY OF RENEWABLE ENERGY SOURCES

VALENTINAS KLEVAS

nova publishers
New York

Library of Congress Cataloging-in-Publication Data

Names: Klevas, Valentinas, editor.
Title: Justification of long-term economic policy of renewable energy sources / Valentinas Klevas, editor.
Description: Hauppauge, New York : Nova Science Publishers, Inc., [2015] | Series: Energy policies, politics and prices | Includes index.
Identifiers: LCCN 2015039934 | ISBN 9781634832038 (hardcover)
Subjects: LCSH: Renewable energy sources--Economic aspects. | Energy development--Economic aspects. | Energy policy--Economic aspects.
Classification: LCC HD9502.A2 J875 2015 | DDC 333.79/4--dc23 LC record available at http://lccn.loc.gov/2015039934

Published by Nova Science Publishers, Inc. † New York

CONTENTS

This research was supported by the Research Council of Lithuania
(IEP-01/2012)

2015

PREFACE

Renewable energy sources (RES) are among the main pillars of sustainable energy development, as the concept of sustainable development envisages the main provision: development that meets the needs of the present without compromising the ability of future generations to meet their energy needs. The idea of the limits of economic growth and finality of energy resources has not appeared until 1980s, so economic evaluation of RES technologies in market conditions is a rather new phenomenon in economic theory. Despite the variety of investigations performed, many questions still remain, which, firstly, should be raised and, secondly, responses should be provided based on a much broader approach and a further perspective than the current ones. Attention should be paid to the fact that renewability means absolute infinity, inexhaustibility of the resources as well as optimistic statements regarding possibilities of energy supply in the future.

A phenomenon of the use of renewable energy in economic terms can be defined as cheapening of RES technologies due to the increase of their demand. These look like exceptions on the general background of the economic development assessment, since the price of any other finite recourses increases with the increase of the demand. However, only renewable energy sources, such as solar, geothermal (aerothermal, hydrothermal), wind, hydro (wave, tidal, etc.) energy, are inexhaustible. Some other resources like biomass depend on the economic activity or the economic process. An excessive demand for biofuels can have very negative consequences in terms of deforestation and environmental pollution. Industrial and household waste constitutes another biomass type assigned to renewables, which could be converted to energy, e.g., landfill gas, biogas from farming, and municipal waste in urban areas. These gases could be used in biodegrading process or in direct burning. Thus, such sources of fuel could be conditionally assigned to renewables. Here, renewability is notional and depends on the conditionally constant accumulation, which could be considered renewability.

Certainly, every type of RES has its advantages and shortages in various aspects. Although some RES, such as solar or geothermal energy, have a scattered character, one should pay attention to a very important aspect: willingness of consumers to use renewables for energy generation for their own needs and not for sale, and thus join energy development process.

In order to equal competition opportunities for those consumers generating energy from RES for their own needs and specific companies, which generate energy for sale, a specific problem should be solved, which is still hardly perceived and rarely analysed in economic theory. This problem is the concept of external public benefit that is difficult to assess and repay for those consumers, who implement this ‘mission’ using RES technologies at their own risks. There are millions of such consumers; thus knowledge transmission is the most important factor here. Huge possibilities to foster RES (especially solar, wind and geothermal energy) implementation process lie in this field. This development trend is rather poorly investigated and implemented with difficulties, though some countries start subsidizing consumers, e.g., a share of investment in the installation of solar collectors is repaid. Social (public) benefits are usually assigned to business companies; however, the profit is the main motivation for businesses rather than meeting social needs of population or meeting energy needs of future generations. Final consumers may be motivated through direct or indirect subsidizing, because namely the consumption of RES creates social benefits. Here, a benefit is understood as a lesser impact on the environment, a solution of some social problems, etc. However, the use of RES faces significant problems in practice. It should be explained why this process is so difficult and complicated, though advantages of using RES are evident.

There is another kind of difficulty. For example, state subsidies for investments are regarded as the simplest form of support that enables rapid popularity of renewable energy technologies if the grant is large enough. The incentive scheme was widely used in various support organizations in many countries in the past. However, it was observed that it had not encouraged project developers to choose economically optimal technologies, and often, unduly expensive plants were installed, operational rates of which were not always justified. The main deficiency of this type of promotion is a lack of connection with the achievement of results. The doubts may arise: isn’t the grant too high? Why and are grants provided reasonably? What is the source of funds, and how long will various RES technologies be funded?

This paper aims to solve the mentioned and many other problems by means of the unanimous theoretical basis, which would certainly allow forming a more united and long-term complex of support measures for RES technologies.

The research was carried out based on the Lithuanian case; however, the solutions and findings can be applied to conditions of different countries, as economic laws are valid for all countries.

The author would like to express gratitude for a significant assistance in this work provided by the researchers of Lithuanian Energy Institute Dr. V. Bobinaite (Chapters 1.2.4; 1.3; 3.1; 3.2.2; 4.1.1; 4.2) and Dr. D. Tarvydas (chapter 1.4). Extremely valuable and significant in the preparation of this monograph were moral support and direct performance of preparatory works, for which the author is grateful to the junior research associate A. Kleviene and the PhD candidate L. Murauskaite.

Valentinas Klevas
Chief Research Associate at
Lithuanian Energy Institute,
Professor at Kaunas University of Technology
Email: Valentinas.Klevas@lei.it
Kaunas, August 2015

INTRODUCTION

Economic research of renewable energy sources (RES) intensified in recent years through the implementation of European Union directives with a particular emphasis on the role of a territorial aspect (cities, districts). Incentives for a wider use of RES are declared both in the European Union and in Lithuanian laws. Adoption of the Directive 2009/28/EC "On the promotion of the use of energy from renewable sources" required the development of a National Renewable Energy Action Plan for 2010-2020 and other documents. This indicates that the development of RES takes an increasingly more important place in the energy policy of the Republic of Lithuania.

However, it may be stated that there is a lack of research, which would answer questions about the new financing methods and economic assumptions that would link the economic policy on the macroeconomic level and would justify benefits of the use of RES technologies state-wide (EC, 2014).

Government support and funding is a very important factor that promotes the adoption of RES technologies in adverse conditions. Government support may take different forms; only some of them are currently used in Lithuania, which basically repeat instruments and financial mechanisms that are used in other countries. Indeed, Structural Funds are one of the most important sources (channels) of funding for RES technologies. Commitments to the promotion of RES under the EU Directives have a significant impact in Lithuania. Government-guaranteed or soft loans allow developers to reduce project risk that is related to renewable energy, and for this reason, provides a lower price of renewable energy. However, the shortage of a soft loan is that it requires considerable financial resources, and their use is difficult to predict in practice. Moreover, these loans have high administrative costs.

It has been observed in the global practice that private sector funding and new forms of organization began to supplement financial support from public funds to RES sector.

It is acknowledged that it is easier to implement an integrated complex of environmental, social and economic policies in cities and districts. Public expenditure and funding from Structural Funds should, where possible, allow achieving more than one policy objective of sustainable development. Sustainable development of energy on the regional level requires institutions to organize and implement support measures. Municipalities can play an important role in promoting assumptions of sustainable energy development, because the local government performs a number of functions in the energy sector, and these opportunities should be used.

Natural monopolies, especially district heating (DH) systems in cities and towns, comprise a special economic and social issue. Although DH companies are regulated by the National Commission for Energy Control and Prices, setting the price cap is just a usual approbation of the performance results. As regards the situation of monopoly and insufficiently oriented pricing, outdated DH may exist without attempting to modernize the technology and reduce the supply.

Development issue of high-tech in Lithuanian energy sector is particularly relevant and largely related to the use of RES. In today's economy, a technological breakthrough would enable one to deal with issues of energy reconstruction. The essential problem and the aim of this study is a unified, long-term, stable, and economically-based formation of economic policy in relation to RES.

This aim determines many tasks, which are grouped into two directions:

a) Assessment of RES impact on various aspects of the country's competitiveness in macroeconomic dimension;
b) Evaluation and justification of existing, forthcoming, and possible to use forms of state support that is based on the influence of RES on the country's economic competitiveness.

In the current economic conditions, utilization rate of RES in Lithuania is excessively low and one-sided, focused on the use of biofuel for many reasons, which are not adequately investigated and identified.

The main issues of economic research and problems are as follows:

- At present, RES are treated as an integrated whole, and all the advantages and disadvantages are classified as generalized, but at the same time separated from specific conditions and the infrastructure that it possible to use.
- Useful effects, which are achieved by the use of RES technologies, are attributed solely to energy producers and suppliers. Energy consumers, which cause the demand, are not evaluated as participants in the process. Accordingly, public support is allocated almost exclusively only to energy suppliers. This effect is scarcely described, and what is more, it is hardly evaluated. Therefore, it is necessary to analyse the assumptions that allow rational assessment of RES benefits in relation to state, towns, energy suppliers, and especially energy consumers.
- The essential issue is the assignment of benefits, as much as possible, particularly to those who invest. For example, RES improve the environmental characteristics of the living environment, but too often, These investments are paid by those who do not get the benefits. Therefore, a territorial aspect of installation of RES technologies becomes an issue; in this context, cities and towns with their DH systems are presently gaining significance, and will continue so in the future.
- There is a scarcity in economic research, which might support the benefit of the use of RES technologies on the state level. Rapid increases in the price of imported fuel and the risk of supply interruption destabilise the economy. This does not create economic assumptions for technological breakthrough of new types of fuels, especially RES, into the market.
- The justification of unified economic policy requires highlighting the main issue: the divide of interests between the producers and users of energy, produced on the basis of RES technologies. On one hand, *producers/suppliers* aim at the higher price and hence the profit. On the other hand, energy *consumers* use this energy or are able to produce energy for their own needs (exceptional cases). The main goal and motive of consumers is to produce energy at the lowest possible price. In the best case, they can realize excess energy, but this requires a set of technical, economic, and legal assumptions and conditions. The use of solar energy demonstrates the disproportion of

such support consequences. Although photovoltaic modules in Lithuania are produced since 1997, there is a legal demand for them in this country. Meanwhile, the Lithuanian production of various photovoltaic modules is exported to Denmark, Germany, England, Netherlands, and even for military industry to Sweden. Lithuanian market demand is currently limited to photovoltaic modules for electronic equipment, yachts, and training devices.

- A great part of the business is the initiative to carry out scientific research, because only in this way, it is possible to survive longer in the international markets. The largest in the Baltic States experimental 149 kW power plant was connected to the distribution network on September 2011. Its purpose is not only to produce electricity from solar energy, but also to carry out investigations of the scientific findings: which of 35 types of photovoltaic devices used in the power plant and which installation methods are the most suitable for Lithuania. Such kind of an innovative organization form is an important factor and example of good practice, because public and private sector cooperation is the most rational future form for the development of RES technologies.

Chapter 1

IMPACT OF DEPLOYMENT OF ADVANCED RENEWABLE ENERGY TECHNOLOGIES ON ECONOMY AND FORMATION OF A MOTIVATION SYSTEM

1.1. HARMONIZATION OF RES SUPPORT SCHEMES

1.1.1. Problems of Promotional Policy Concerning Sustainable Energy Development

The concept of sustainable energy development cannot be separated from the understanding of the additional positive socio-economic effect of sustainable energy projects (energy efficiency measures, use of renewable energy sources). Implementation of sustainable energy projects has positive impact on security of energy supply; it provides financial economies and improved comfort and has multiplier effect for new jobs, involving small and medium size enterprises. Therefore, implementation of sustainable energy projects may create an external positive effect concerning increased energy security and other development goals (reduction of unemployment, reduction of environmental impact, etc.). The examples of EU-15 countries implementing their sustainable energy development and climate change mitigation policies on local level can be successfully applied in Lithuania.

In the last decade, EU policy has been changed mainly through supporting a suitable policy of distribution and use of energy with the main objectives of ensuring the reduction of the consumption of resources and environmental degradation. This political orientation has created and underlined a wider role of the local and regional systems in the implementation of European energy policy. However, EU policies together with national policies are not completely in harmony and need to be integrated on a regional and local level. Furthermore, a significant number of policy tools (most importantly the programmes) with different levels of autonomy in design and implementation need to be brought together on the local level. Special attention is given to the role of the regions and local areas in the implementation of EU energy objectives and policies (Antinucci et al., 2002).

These objectives have been strongly pursued and integrated in different EU policy and programme documents, the main document of which is the "Action plan to improve energy efficiency in the European Community." The Action plan mainly aims to integrate and give coherence to a combination of several policy instruments as well as ensure that they reinforce each other. The reason can be found in the pressing need to renew commitment both on the Community and Member State level to promote energy efficiency more actively. Policy instruments are designed to complement and reinforce Member State activity in this area and not to duplicate them.

The proposed actions of the Action plan in this context consist of:

- Measures designed to integrate the energy efficiency dimension with Community non-energy policies and programmes;
- Measures to strengthen and expand existing energy efficiency policies and measures, and new policies and measures.

There is an important role foreseen for the regions and local areas in the take up of sustainable energy policies. This role, nevertheless, is only envisaged as a stronger move towards integration of different and co-existing policies, programmes and responsibilities that have yet to be carried out.

Recently, a number of critical remarks concerning the measures for promotion of implementation of renewable energy technologies and production of energy from renewable energy sources (RES) have increased. One of the principal shortcomings of the promotional measures, which worry both the scientific community and practitioners almost in all countries, consists of the following: measures for the promotion of RES energy frequently contradict market rules and distort the market (Ciarreta et al., 2011; European Commission, 2014; EURELECTRIC, 2014). Moreover, they are used to implement short-term objectives and do not create any unified (harmonized) and long-term oriented programmes. The newest EU technical document respecting creation of support schemes for promotion of renewable energy technologies indicates the necessity to revise the currently existing RES support measures, which provoke market failures. "With growing renewable shares, poor design and implementation of public intervention has led to unnecessary distortions with regards to energy production, trade and investment in renewables. This raises the cost of the promotion of renewables and risks hampering both the further growth of renewables and the completion of the internal electricity market" (EC, 2013). Certainly, good practices exist. An example of good practice is the economic policy implemented in Denmark. It binds the main elements ensuring success of economic policy into a single unit: a) a long-term (up to 2050) perspective; b) an energy strategy, which is linked to a well-rounded support policy for RES technology implementation; c) external cost, which emerges due to the combustion of fossil fuels, is included into the energy price as a part of taxes and, thus, improves competitive positions of RES technologies in the market; d) understanding that due to the natural reasons, RES technologies cannot ensure total capacity for supply of energy demanded; they are used in a combination with fossil fuel technologies in the scenarios; e) infrastructure, such as district heating system, is used for the implementation of RES technologies; f) natural monopoly is used as a powerful measure, which creates possibility for a diversification of fuels and the stability of prices.

An EU document proclaims that support for RES technologies is often unbalanced. "Renewable energy producers have faced a range of frequent changes to support schemes. Change is constant. In the renewables sector, the regulatory risk that comes with such changes has a direct impact on capital financing costs, the costs of project development and therefore with the whole process of developing renewables" (EC, 2013). However, one has to take into account that development of renewable energy policy is a complex system, where a balance of three aspects of sustainability – environment, economy and social life – is needed. Contemporary generation has a challenge to cope with issues without leaving them to other generation to solve. A good performance and well-balanced RES policy needs efforts from different actors and stakeholders. It is believed that although each country has different starting conditions, namely, developed technical infrastructure and own volumes of energy resources, still, development of energy from RES should be shaped on the basis of a long-term macroeconomic policy. This means that among other actions to be taken, a good and scientifically based theoretical background is needed, i.e., both indicators reflecting economic development and a method for assessment of impact of the certain type of energy from RES on these indicators are required. This monograph will deal with these issues.

Lithuanian solar and wind power sector was chosen for the analysis. The motivation for the selection of the country and type of RES energy originated from the peculiarities and uniqueness of energy sector and its regulation (including support schemes) in the country. Lithuania as a case study was examined considering the fact that energy from RES is supported through time-varying support schemes (subsidies, feed-in tariffs, fiscal measures and combination of them), which basically repeat instruments and financial mechanisms that are used in foreign countries without deeper understanding of cost and benefits of such support scheme. Moreover, it was observed that RES support scheme applied in the country does not encourage project developers to choose economically optimal technologies, and often unreasonably expensive plants are installed, operational rates of which are not always justified. The main deficiency of this type of promotion is a lack of connection with the achievement of results. Thus, this paper will introduce the theoretical background of the economically efficient RES support scheme.

Wind power sector was selected considering the benefits that wind power provides. This benefit is so sizable that it could easily justify a larger share of wind energy in most European countries, even if wind were more expensive per kWh than other forms of power generation. This risk reduction from wind energy is presently not accounted for by standard methods for calculating the cost of energy, which have been used by public authorities for more than a century. Quite the contrary, current calculation methods blatantly favour the use of high-risk options for power generation (EWEA, 2009).

Although a substantial research has been done in respect to RES, there still remain some uncertainties, which need to be identified and solved by energy policy makers through determining the reliable directions on how to develop RES sector in future. Intuitively, it is perceived that development of a harmonized economic policy is necessary for RES sector development; however, it is apparent that mistakes are still made, so they need to be recognized and corrected. The newest EU document proclaims that support for RES technologies is often unbalanced.

Namely, in the less developed regions, investments under the auspices of structural funding should concentrate on the following:

- Investment in renewable energy sources favours the development of local resources and contributes to the reduction of dependency on energy imports, while at the same time creating new local job opportunities. Targeted aid could be used to promote the acquisition of renewable energy equipment by SMEs, for instance, in the tourism sector, where there is a considerable potential for the use of renewable energies.

The aim of the Programme "Intelligent Energy for Europe" is to accelerate the market penetration of new technologies in the fields of RES and energy efficiency by creating a favourable legislative framework and laying down objectives for the EU and for the Member States. This new approach also involves strengthening of Community action vis à vis the main stakeholders including national, regional and local agencies and any other organisation contributing to the Community effort in line with the changes proposed in the White Paper on European Governance, which aims at achieving among other things greater participation and openness between the main players concerned by establishing a more systematic dialogue and interaction between the Community and the Member States, the regional and local government bodies and civil societies. The main objectives are:

1. To make it possible to develop and implement a medium and long-term energy strategy, which is compatible with the aims of the EU strategy on sustainable development and the legislative measures necessary to achieve a follow-up and monitor the strategic objectives.
2. To bring about a genuine change at Community level by raising awareness among the main players so that they move towards efficient and intelligent production and consumption schemes based on solid, sustainable foundations.
3. To promote exchanges of experience and know how.
4. To support activities intended to boost investment in emerging technologies already present in the market.
5. To encourage the spread of best practices, not least in the educational system.
6. To increase transparency, coherence and coordination of all the activities, to promote activities which combine various fields, and to create an effective link between action under the programmes and activities carried out under Community policies, especially the strategy for sustainable development and the European Climate Change Programme, development cooperation, policies to strengthen economic and social cohesion, especially in the context of an enlarged

Union, and the structural measures of the ERDF in the energy field, as well as the Member States' initiatives.

EURELECTRIC, the Union of Electricity Industry, has published the document, which states that the primary and fundamental target of power market liberalization aiming to increase the competitiveness of Europe has not served this purpose. The main reason for that is that there are many non-related policies, initiatives and directives across Europe, and the achievement of three main purposes in parallel is complicated, because different economic and legal tools, measures and mechanisms, sometimes conflicting with each other, are used. EURELECTRIC has also criticized governments for providing "substantial and unlimited" subsidies for RES. The association believes that sustainable growth in this area can only be achieved through a regulatory framework, which complies with the rules of a competitive market. Maintaining that support schemes for RES have to be market based, capital efficient and designed to avoid market distortions, EURELECTRIC calls for harmonization of RES support schemes (EURELECTRIC, 2004).

1.1.2. Sustainability Assessment of the RES Project Implementation

Applied welfare analysis presents a classical approach to solving the problems above. An efficient resource allocation occurs in the way energy is priced. At any time, the price of the energy resource should reflect its marginal social cost. This implies that any divergence between social and private cost arising as a consequence of the market failures (for example, public goods or externalities) should be corrected by internalizing such external costs. To assure the equivalence of price and marginal costs, a competitive market framework is required. Monopoly elements are to be eliminated if prices are not equal to marginal costs. Applied welfare analysis allows quantifying the welfare loss associated with any given market failure. This enables policy makers to focus on the market distortions. But even though a policy change leads to an aggregate welfare gain, certain groups may suffer deleterious income distribution loss. Again this implies a need to create welfare system to redress burdens imposed on the poor as a consequence of an efficiency generating policy change.

The main problem according to the theoretical concept is as follows: market consists of economic agents operating under perfect competition conditions. All market-based mechanisms provide for efficient resource allocation just under perfect competition, which is not available in reality. “An urge for a liberalization of global trade has been sweeping anonymous over the world without any public debate and without even asking what are the benefits. Indeed a market economy is excellent on a certain scale. Today it is no longer a free market. From a democratically controlled market we have then moved to a market controlled by the big market forces themselves. The criteria behind the economic theories of the advantages of a free market no longer exist” (Norgard, 2001).

It is even worse that methodology of cost-benefit analysis, where positive or negative effects of increasing energy supply security, social, environmental, economic impacts are treated not as externalities, but as normal commodities. The main difficulties arise when atmospheric pollution and external costs are treated as normal commodities. The following theoretical considerations reflect this problem: “Critical policy question is not whether pollution exists but rather whether it occurs at socially optimal level. The economic approach to pollution control is to view clean air and water as commodities like gasoline and air conditioning, to be provided according to the standard criteria applied to all goods” (Griffin and Steele, 1986).

This problem is upgraded to the problem of national accounting when growth of GDP indicates constant progress. However, the investigation of economic growth and valuation of environment correct the optimistic result provided by constant national income and GDP growth in developed economies (Van Ireland et al., 2001).

According to the economic theory, the problems related to achievement of sustainable development goals can be solved by balancing marginal social costs and marginal social benefits. This theoretical concept opens the pathway to “sustainable development” not restricting the growing fierce demands of consuming society. It is truth that the growth of GDP is pumped by growing consumer demands. A major shortcoming of the above-mentioned attempts to adjust the conventional GDP might be that they still aim at just one index to encompass the diversity of what makes up a life and a culture.

Correction GDP ***As Welfare Indicator***

Attempts have been made to correct the GDP. One of them is a Genuine Progress Indicator, GPI (Norgard, 2001). With some rather modest and very reasonable adjustments of GDP, the trend is a decline since around 1965, and at the present level, it is lower than ever in more than 45 years. One of the changes is that for this GPI, the cost of restoring environmental damage is counted as negative rather than as positive, as is the case for the GDP. Despite the indication that with the present trend people are getting worse off year by year, the politicians in the USA and other affluent countries are dedicated to pursuing the same conventional target, which means the downward trend for GPI.

Another alternative to GDP was the Index of Sustainable Economic Welfare, ISEW, which showed similar trends. While these attempts clearly show the incompetence of GDP, none of them claim to have found *the* right indicator.

Sustainable Value Concept

Value-oriented methodology to assess the sustainable performance of companies and other economic entities is called Sustainable Value. Figge and Hahn (2004) propose a new approach to measure corporate contributions to sustainability called Sustainable Value Added (SVA). Value is created whenever benefits exceed costs. Current approaches to measure corporate sustainable performance take into account external costs caused by environmental and social damage or focus on the ratio between value creation and resource consumption. As this paper will show, it is more promising to develop sustainable measures based on opportunity costs. Sustainable Value Added is that kind of a measure. It shows that a significantly greater value is created, because a company is more efficient than a benchmark, and because the resources are allocated to the company and not to benchmark companies. The concept of strong sustainability requires that each form of capital is kept constant. As Sustainable Value Added is inspired by strong sustainability, it measures whether a company creates extra value while ensuring that every environmental and social impact is in total constant. Therefore, it takes into account corporate eco- and social efficiency as well as the absolute level of environmental and social resource consumption (eco- and social effectiveness). As a result, Sustainable Value Added considers simultaneously economic, environmental and social aspects.

From a theoretical point of view, these concepts provide very powerful measures of corporate contribution to sustainability, because they thoroughly translate the requirements of the constant capital rule on the macro level into measures for the micro level. However, there are a number of shortcomings and problems. Here VA of a company is defined as the residual value that remains after the turnover has been reduced by the cost of goods and services purchased by the company. SVA represents the aggregate of all energy and material flows under consideration induced by economic activity weighted by their relative harmfulness to the environment. This aggregation is done by classifying every emission according to its contributions to different environmental problems, which are then weighted by their relative harmfulness (for methodological details see, e.g., Heijungs et al., 1992). Eco-efficiency describes a degree to which a company uses environmental resources relative to its economics.

Sustainable Value Added considers efficiency *and* effectiveness of all three dimensions of sustainability. Because it shows the amount of value created while ensuring a constant environmental and social performance, Sustainable Value Added is based on the paradigm of strong sustainability. Put differently, Sustainable Value Added represents in monetary terms the extra value created by a company adjusted for all changes in eco- and social effectiveness. Compensation in this context does not mean paying victims for accepting external effects, which would imply weak sustainability and thus substitutability of different forms of capital. Compensation for calculating Sustainable Value Added refers to a purely allocative question and means paying other, less eco- or social-efficient users of resources to reduce exactly the environmental and/or social impacts in question. This results in a constant overall level of eco- and social effectiveness. As only identical environmental or social impacts are considered in that kind of compensation, substitutability between different forms of capital does not matter. Sustainable Value Added is thus based on the paradigm of strong sustainability.

Sustainable Value Added provides a new approach to measure corporate contributions to sustainability. However, this approach focuses on method development rather than on the implementation of the method. Therefore, there is a need for further research in two directions: on the one hand, empirical surveys are needed that apply the method to a wide range of companies of different industries in order to examine both the practicability of the method as well as the results it elicits in real world terms. On the other hand, future studies should investigate the institutional and organisational conditions and arrangements of some kind of trading or market scheme to implement Sustainable Value Added. In addition, Sustainable Value Added is limited to the effect that it does not indicate, if a company is sustainable. It shows, however, how much a company has contributed to more sustainability. This contribution can be expressed in either economic and environmental, or social terms. When expressed in economic terms, Sustainable Value Added expresses in absolute monetary terms the sustainable performance of the company relative to a benchmark. By establishing such a micro-macro link, it expresses by how much a company contributes to more sustainability on the benchmark level.

Multi-Criteria Analysis

The multi-criteria analysis can also be applied for the selection of sustainable energy projects to be supported, e.g., financed from Structural Funds. The criteria for the selection of sustainable energy project having positive impact on sustainable regional development are: efficiency, impact on environment expressed in terms of decreasing external costs, impact on job market, equity, technological innovation and security of energy supply. When policy-makers and stakeholders are asked to select the best project, they have to find a solution that gives the best outcome in terms of the criteria mentioned. However, it is difficult to identify a project that outperforms others in terms of efficiency, impact on environment, job market, security of energy supply, equity, and technological innovation. The choice of the project will require some trade-offs among these criteria (Hanley et al., 2001).

A conjoint choice analysis can help investigating how policy-makers and stakeholders trade-off the criteria when designing a policy for the internalization of the externalities, selection of investment projects for financing, etc. Conjoint analysis and other stated-preference (SP) techniques have recently emerged as a complement to revealed-preference (RP) techniques. While RP evaluate economic agents' behaviours in real markets, SP involve choice responses evoked in hypothetical markets. The interest in hypothetical behaviours in economics arises from different reasons, such as the necessity to investigate economic agents' preferences for new policies that might be implemented, for evaluating goods that are not traded in real economic markets. This method is applied when it is not possible to estimate agents' preferences using revealed preferences. During the choice experiments, individuals or policy-makers are asked to choose between two or more projects described by a set of attributes or criteria (environmental impact or external costs of the project, impact on security of supply, equity, job market, etc.). The levels of the attributes vary across the projects, so that respondents trade them off. One of those attributes is usually a cost amount (for example, external costs), which allows the computation of marginal prices of the attributes, which are difficult to evaluate in monetary terms (for example, impact on security of supply (Louviere et al., 2000).

In a single choice experiment exercise, researchers learn only which alternative is the most preferred, but the result of the exercise does not tell anything about the preferences for the options that have not been chosen. A single choice experiment exercise does not offer a complete preference ordering. Therefore, if researchers want to know a complete ordering of preferences, it is necessary either to ask a respondent to do many choice exercises, or to survey more respondents varying the levels of the attributes (Markandya, Longo, 2005).

Strategic Environmental Assessment

Another measure for sustainability assessment of energy projects is Strategic Environmental Assessment (SEA). It allows for integrating of environmental considerations into economic development, which has received considerable nominal support amongst governmental and non-governmental groups in the strategic environmental assessment of development-related policies and programmes. SEA Directive 2001/42/EC was adopted on 31 May 2001. A wide range of plans and programmes begun after 21 July 2004 now require an environmental assessment. The purpose of the SEA-Directive is to ensure that environmental consequences of certain plans and programmes are identified and assessed during their preparation and before their adoption. The public and environmental authorities can give their opinion and all results are integrated and taken into account in the course of the planning procedure. After the adoption of the plan or programme, the public is informed about the decision and the way in which it was made. In the case of likely transboundary significant effects, the affected Member State and its public are informed and have the possibility to make comments, which are also integrated into the national decision-making process. It is expected that SEA contributes to a more transparent planning by involving the public and by integrating environmental considerations. This will help to achieve the goal of sustainable development.

However, the culture of economic development is still resistant to the broader integration of environmental concerns (ExternE, 1995). The integration of strategic impact assessment procedure for project appraisal and selection should be critical in use of Structural Funds. As far as SEA is concerned, the experience gained in the 1990s has shown that it has been a considerable advance in practice both procedurally and methodologically at the level of programmes and plans (Treivel, Partidario, 1996; Gouldson, Peter, 2000). The most important issue is the change in the culture of economic development, and a reinterpretation of the usual quality-of-life measures could promote a better integration of environmental and economic strategies: not only through using environmental appraisal criteria, but by greening local economies through promoting certain projects. The process of sustainability appraisal could lead to an adjustment of economic development programmes to deliver sustainable energy projects.

All three approaches described above can be successfully used for the integration of sustainable energy projects into regional development programmes.

The Idea of Eco Villages and Green Settlements

The idea was formed in the UN Habitat declaration in 1976 year. It had become rational and effective in various aspects, where one of the significant aspects was the use of energy. Following the principles of Agenda 21 accepted at Rio Summit in 1992, the baseline of sustainable development for settlements was accepted, and finally the essential principles were accepted at Habitat-II UN conference in Istanbul in 1996.

Declaration for sustainable development of settlements states that cities and settlements were and are the incubators of our civilization and generators of growth, which have accelerated the evolution of knowledge, culture and traditions, as well industry and trade. Proper planned and controlled urban settlements create favourable perspectives for human development and protection for global nature resources.

The movement of "eco villages or "green" settlements is driven by local initiatives, local community and group leaders to pursue the main objectives: to create better, healthier, eco-friendlier lifestyle conditions, fostering social and community lifestyle, and encourage the development of ecological permaculture and nature economy.

According to the information of International organization "Global Eco village network," there are more than 40 thousand ecological villages around the world. A movement of ecological villages is usually driven by local community initiatives; however, more often it has been supported by governmental subsidies. For example, British Government has accepted and supported the construction and development programme of four new eco-villages (LR Ministry of Agriculture, 2010). However, the main problem now is that the supply of energy resources will still exists in the rural areas in the future. The main question altogether emphasising the actuality of this scientific work is to connect various research results into a united and practically applied model of "green" settlements covering criteria of sustainable development.

The use of green energy is one of the priority issues in the projects of "green" settlements developed by local communities; however, it forms a separate group, where thermal and electrical energy production and consumption are considered. The main aspect of formation of the energy economy model of "green" settlements is the optimal use of RES and effective energy consumption. Since now there are no directives or legal norms regulating the development of “green” settlements, the development of such settlements is considered as local initiatives driven mostly by local leaders and using the bottom-up approach.

The energy model of “green” settlements is a formation of electric and thermal energy production and consumption scenarios, where the use of local and renewable resources is considered. The formation of energy model for “green” settlements assumes that the share of RES should be no less than that indicated in the Directive for promotion of use of renewable energy sources. However, the perspective development of “green” settlements cannot be stopped by 2020. Preferably in the optimal case, an assumption is accepted that the model of “green” settlements in the future must be based on 100% of the use of RES. Yet, this technical target is just a guideline, because other factors like economic, social and environmental also need to be considered. In this way, the development of “green” settlements is a continuous process without an end like in the cycles of nature, where one form changes the other. Consequently, it is important to scientifically justify the systematic use of renewable resources, the reserves and waste management using resources flow analysis. It is necessary to develop a secondary use of materials (reuse), where toxic waste is involved in the technical (anthropogenic) and biological (ecosphere) cycle.

“Green” settlement is related primarily to the understanding of ecosystem approach and holistic approach to the formation of ecologic lifestyle. Living organisms and their environmental interactions in the nature are interdependent processes that work together ensuring the life and convenience of the system (Palojarvi 2013). Therefore, the development of the model of “green” settlements is intended to preserve the entity of the nature and to emphasize the dependent position for each of its parts and for human as well (Coriander, 2007).

The principle of ecological movement can be expressed by the quote of scientist Commoner “think globally, act locally.” Commoner noticed that the progress of solving environmental problems can be achieved quicker if the principles of our activities and the function of natural law systems are harmonized. This was the main principle of the development of “green” settlements. However, the energy economy of green settlements duplicates the processes in the nature, which is a perfect example of waste-free technology. The amount of waste generated in the energy economy can be reduced by increasing the efficiency of use of primary energy resources, promoting the use of RES, choosing eco-friendly products and materials whose components are less harmful to the environment. The negative impact to environment can be mitigated through more efficient use of renewable energy resources in the energy economy developed under the circular economy model and bioregional development principles. The use of RES in the region can create unique environment for resource utilization (Birekland, 2008). However, the exploitation of energy resources must not exceed the resilience limits of nature, otherwise RES can be treated as non-renewable resources (Ciegis, 2009). The development of regional energy economy, according to principles of circular economy, is encouraging the use of a wide range of local and renewable resources (Pledger, 2013). The concept of circular economy and the development approach from cradle to cradle are investigated by Wijkman, Rockstrom, Walter Stahel, Braungart, McDonough.

"Green" settlement is like a living laboratory, where ecological innovations are created and the solutions are accepted ensuring the sustainable development in the sector of civil engineering, energy use, waste and waste water management, transport organization, agricultural and food production, the culture of consumption. Therefore, a process network analysis method (PNS) is applied for research in this "laboratory." PNS method is associated with the creation of an algorithm of branch-and-bound method, which helps to conceptualize the present situation, forecast options of feasible decisions, support identification of the kind of information needed in order to evaluate the consequences of decision-making.

1.2. The Concept of Usefulness of Renewable Energy Sources and Evaluation of Sustainability of RES Projects

1.2.1. Object of the Research

Firstly, the task is to determine the object of the research. RES is a too wide object, and it is not feasible to analyse all possible types of renewable energy sources, their application, technologies available for use and, particularly, their usefulness (utility). Moreover, utilization dynamics of each type of RES are different. We have in mind the reduction of cost of technology with the time. Thus, the initial objective is to transpose all discussions and considerations into a certain plane, which is defined by unanimous rules and economic laws, and to find a common denominator of different opinions. This means that single facts should be "bind" into a logical chain of actions and consequences, i.e., into the chain, which is based on economic laws and is called theory.

Figure 1.1 shows RES classification scheme, which displays social utility categories under the particular type of RES.

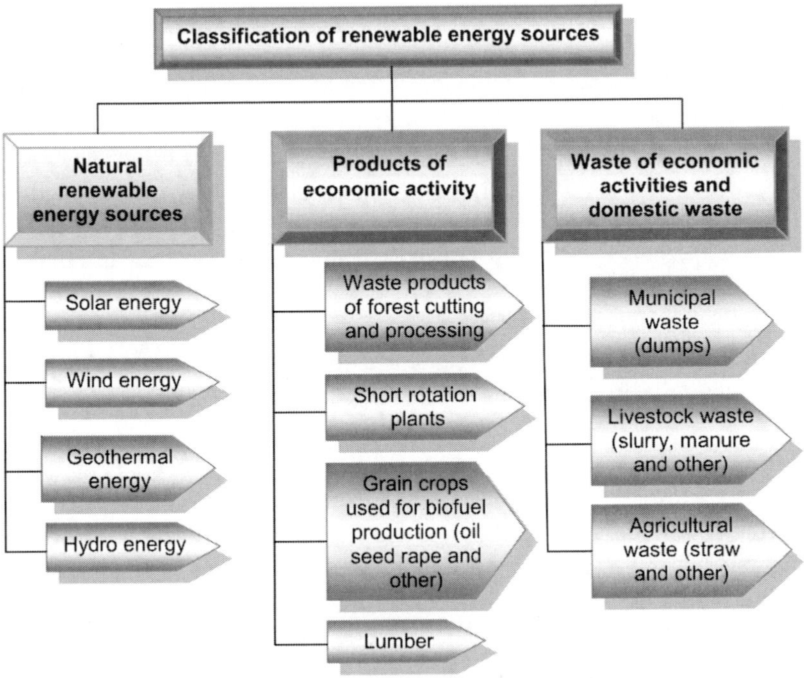

Figure 1.1. Principal scheme of RES classification.

The usefulness of renewable energy sources differs from that of oil and gas. The main difference is that renewable energy sources ensure that future generations will be supplied with energy sources. This substantial advantage is not methodologically assessed. But this is one of the fundamental accents of concept of sustainable energy development. Renewal, as an exclusive feature of these energy sources, is important for classification. RES are grouped into three main types. The first or the basic type is a phenomenon existing in the nature are renewable energy sources, which during a long time, however, gradually can substitute exhaustive energy resources, for example, oil and gas. This is solar, wind and geothermal energy.

The second type is energy resources received as a result of economic activity. Their conditional renewal depends on continuity of the economic activity. Moreover, in all cases, one has to bear in mind multifunctional resources, for example, utilization of lumber.

The third type is utilization of waste from economic activity and municipal waste. When RES technologies are employed, then waste utilization is a good tool for solving environmental problems. The employment for energetic needs in this case is rather limited.

RES support scheme should be differentiated *in accordance to* social benefits it can provide:

- natural renewable energy sources mean that their utilization is not exhaustive; this gives possibility for future generations to self-provide with energy resources; however, it is necessary to bear in mind that renewal is different;
- in the sense of circulation of natural processes, they mean that technological progress is oriented towards harmonization of human activity with circulation of natural processes.

In the framework of this programme, the research object is natural RES, which are currently simply used, but wide utilization of them gives an opportunity for future generations to ensure provision of energy resources. Unanimity of motivation of RES suppliers and consumers is the essential problem both from scientific and practical point of view. Under the normal market conditions, this does not cause any problems; however, RES is a specific resource. The problem is that business is interested in high purchase price of energy and establishment of infrastructure by using state funds. Industry push emerged in some market segments (for example, small-scale hydro energy and wind energy).

Meanwhile, consumers seek for lower prices, and this aim contradicts the aim of energy suppliers. Therefore, a question arises: how to balance the demand with the supply, or who pays premiums to tariffs. Besides, it is necessary to clearly know on what ground a business has to be supported.

In a theoretical context, the task is to prepare a theoretically-reasoned systematic approach, which is dedicated to motivation of RES utilization. One of the main roles of science is to provide information, which allows better shaping development policy in the decision-making process. For this, it is essential to basically improve scientific understanding, long-term assessment, which is reasoned on scientific research, and to improve scientific State's competence by new research scheme, which would allow receiving answers to questions that remain in shadow when RES support schemes are transposed from international practice. When transposing RES support schemes from international practice, State's economic status in society's evolution process is considered insufficiently.

1.2.2. Reasons of Necessity to Support Adaptation of RES Technologies

Sustainable energy development means energy production and consumption, which ensures State's and, although this sounds pretentiously, mankind's long-term aims consistent with social, economic and environmental dimensions. In other words, this is reliable energy supply at an affordable price and causing as low negative impact on environment as possible. The most important dimension of sustainable energy development is energy supply reliability or energy security, i.e., possibility to use energy in its various forms at any time, in sufficient amounts and at affordable prices. Energy security plays a crucial role in development of mankind since rough organic fuel, on which energy systems of many countries are based, distribution in the world influence on the possibilities of single countries to develop their economies and ensure wealth of population. Although markets play the most important role in ensuring reliable energy supply, because of external cost related to reliability of supply, energy markets cannot ensure energy supply reliability. Therefore, additional measures and agreements of Government and international organizations are required, seeking to ensure energy security.

Renewable energy sources are more evenly distributed in the world compared to resources of organic fuel; however, a range of factors, such as competing use of land, environmental limitations in the locality, time of utilization of wind and solar energy, significantly limit the economic potential of renewable energy sources. Seeking to improve utilization of renewable energy sources, support for renewable energy sources is required. Two important energy market failures prevent development of utilization of renewable energy sources. These are 1) external energy supply cost emerging due to pollution and 2) external positive benefit of renewable energy sources as a public good. This benefit is related to adaptation of new technologies, their development and positive influence on environment. Thus, seeking to overcome aforementioned market failures and other market barriers, hindering the development of renewable energy sources, it is required to subordinate market mechanisms, reduce renewable energy adaptation cost, increase competitiveness in the markets and promote development. Here subsidies for renewable energy sources can be applied, or obligatory purchase quotas for energy produced from renewable energy sources can be introduced.

Markets also do not ensure financing of long-term research and development. New energy technologies are permanently impacted by barriers in all states of energy innovation chain (from research and development to dispersion of demonstration projects, information and technologies).

The main market failures, preventing implementation of the aims of sustainable energy development, are the following:

- energy markets are not capable to ensure energy supply reliability;
- markets do not assess environment and social development dimensions;
- markets do not ensure long-term research & development, increase of energy efficiency & renewable energy sources, development and implementation of new energy technologies;
- markets do not guarantee that poorest people will be self-sufficient with energy.

Although market, managed by competitive forces, ensures optimal distribution of resources better than administrative systems, it cannot assess social and environmental cost of energy production and consumption; market also does not guarantee sustainable energy development. Policies for 1) energy supply reliability, 2) energy efficiency measures and 3) new energy technologies in the market are required. Besides, special State support measures, which would ensure energy supply to the most vulnerable part of the population, are required. Thus, focused Governmental policy and efficient regulation are essential in areas where market cannot ensure important public benefits. Seeking to reduce market distortions and market shortcomings, in case external energy production cost is not assessed when allocating resources, it is required to implement such measures as 1) removal of subsidies to various types of traditional energy, 2) integration of social cost into the energy price in the shape of levies or emission trading and 3) to implement flexible market measures ensuring energy supply reliability. Such regulation measures show that it is essential to acknowledge that restructuring and liberalization of energy market will not ensure sustainable energy development, although they help regulating monopolies and embedding competition in energy markets.

However, firstly, an approach, which will disclose on what principles assessment method is built, and how research should be organized, is required.

Figure 1.2 shows distribution of benefits concerning RES utilization.

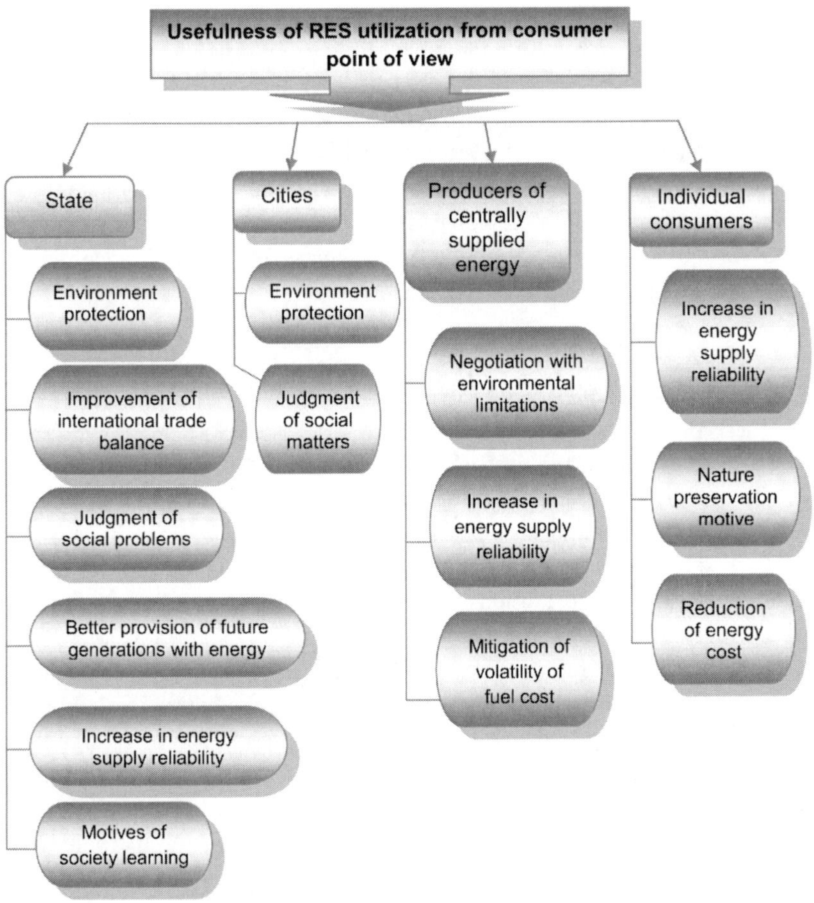

Figure 1.2. Usefulness of utilization of renewable energy sources from the consumer point of view.

An especially important question, which requires solution, is the following: What direct and indirect benefits could specific types of renewable energy sources have?

Many countries view renewables as a way of promoting development of small and local businesses in selected areas and diversifying supply patterns at the regional level. RES (including biomass, solar, wind, geothermal and hydropower) have the potential to provide energy services with zero or almost zero emissions of both air pollutants and greenhouse gases. In addition, the greenhouse gas abatement due to a more intensive use of RES would contribute to the achievement of the EU target, which is related to climate change; this being the fourth target in its energy strategy (Cansino et al., 2011).

However, integration of sustainable energy projects, e.g., RES technologies, may be successful after a reliable methodological assessment of positive effect of such projects for solving social, economic, rural development problems in the context of regional development. The scientific issue is to define the economic background for policies and measures aiming at sustainable energy development (Klevas, 2010).

1.2.3. Theoretical Explanation

Most investors, while choosing places for their investments, usually analyse strategic plans, and these are critical factors for choosing a place for investments or at least choosing possible variants.

The need to support RES is defined by the shortcomings of the market, which create unequal conditions for comparability and competitiveness of various types of fuel and energy.

There are several reasons why in Lithuania, as in many other countries, solar energy is used distantly. It is hypothesized that the current situation does not only exist because of the lack of economic support. The more significant reasons are a lack of organisation, gaps of public education and dissemination of knowledge in society, also too high expectations of getting the support from the state. Generally, it is operated on standard opinions that are partially right, but more important factors, which vary in time, are not considered. Knowledge is usually concentrated and presented in a narrow approach, when some facts are emphasised and some are suppressed. For example, it is assumed that economic potential of energy made from solar thermal energy is poor because of the expensive technologies and inappropriate climatic conditions. This fact is easily denied by the countries with similar climatic conditions, where the development of solar energy is vigorously and rapidly increasing next to other types of RES.

A breakthrough was noticed even in Lithuania in 2008–2011, especially after the feed-in tariffs for the electricity, produced in solar power plants and supplied to networks, were announced. Also there are many examples that even without the high support from the state, numbers of consumers install solar energy devices. However, these facts are outside the official statistics yet.

The opportunity of the use of solar energy and the potentially huge consumer market are not getting enough attention in the Government documents. Therefore, in order for this type of energy to be used to a greater extent in Lithuania, effective support from the Government forming the basis for a competitive solar energy price is necessary. However, the support does not have to be in direct subsidies or other financial forms that are paid from budget or by consumers. "Cheapness" or "expensiveness" is more a psychological concept varying in time. Moreover, both cheapness and expensiveness are temporary conceptions that strongly depend on the extent of technology implementation (ESTIF, 2011). For example, in Europe, an evident trend of decreasing costs of solar hot water systems is observed. The price decreased by 40% from 1995 to 2006 and reached about 870-1015 €/kW; the forecast is that the price will drop to 435-465 €/kW by 2025. Prices in Lithuania are about 725-900 €/kW at the moment.

The other standard opinion concerns the "cheapness" of biofuel, which seemingly should rescue the heat sector. The prime cost of the heat increased by about 30% in all municipalities (except Lazdijai) in the last decade. It is maintained that this increase of price was determined by the rises in cost of fuel (the price of natural gas was 104 €/toe in 2000 and 368 €/toe in 2010; of the regional market. Unfortunately, it has to be admitted that there is a one-direction movement to the Scandinavian countries.

From the point of view of RES supply, support tools for technology creators and suppliers means compensation of social RES production costs, as traditional resources are not evaluated correctly because of market errors.

From the approach to RES demand, one should investigate correct evaluation of social benefits, which may show those advantages that are not seen in investment solutions, e.g., inexhaustibility and possibility to ensure energy resources for future generations. RES is a wide group of energy resources, and the assessment of these resources depends on the potential, secure and sufficient supply, environmental and renewable character, as well as the impact on solution of social problems.

Economic theory explains the main reason why there is not enough demand in the market for some products and services, although they are very beneficial for the society. The reason is that the external benefit of the product or service generally is not included in the direct market operations and in prices (Klevas et al., 2010). It is a particularly important moment in terms of

RES. Demand of RES technologies does not indicate the benefit of the society (social benefit), which is gained by using the energy produced from RES. The benefit is dispersed between different users, which can be contemporary people, future generations, cities, companies. People, even if they do not participate in the market transactions, sometimes can get some product benefit for free. The external environmental benefit of using RES is probably the best example of external benefit. Therefore, due to the RES use for energy production, the negative environmental influence is being reduced, organic fuel resources are being economized, new technologies are being installed, etc. Figure 1.3 shows the "green" (made from renewable energy resources) curve of electricity supply S that involves the expenditures of producing the "green" energy. The main expenditures cover machinery and equipment, new technologies, employee's wages and other expenditures for the training of staff, qualification improvement courses, etc.

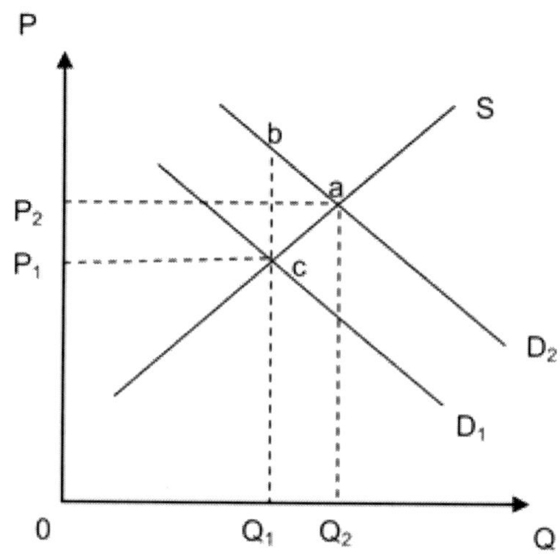

Figure 1.3. Probable influence of external benefit evaluation on RES demand.

If the benefit from the "green" energy consumption is unknown or unvalued, consumers will choose the volume of purchased energy by demand curve D_1. "Green" electricity will be bought in smaller amounts than it would be economically reasonable: quantity Q_1 instead of Q_2. However, when buying "green" energy produced by RES, the pollution for all residents reduces. Everybody avoids diseases and other negative consequences that emerge from the polluted environment. This benefit is not included in demand D_1. The curve of demand should be D_2, where the external benefit is evaluated. The main problem is to include the benefit into the expression of economic indices.

Marginal benefit of every additional unit of electricity from the quantity of Q_1 to Q_2 exceeds their marginal production costs. External benefit that is shown in the triangle “abc” is not realized and represents the inefficiency of the market. It is notable that the benefit is known and declared, but this knowledge is not converted to concrete practical calculating mechanisms. However, conclusive calculations are not enough. It is necessary to significantly increase structural changes in order to initiate the process of RES development. Especially important is that knowledge becomes the main driving force of this process.

Without the state intervention, competitive market will produce “green” energy in point “c,” where market supply curve S crosses demand curve D_1. Market equilibrium price P_1 will be at this point, and the quantity of produced “green” energy will reach Q_1. However, if not only private but also external benefit would be evaluated, the new equilibrium price would be $P_{2,}$ and the quantity of produced “green” energy would increase to $Q_{2,}$ where the marginal benefit is equal to the marginal costs. Social benefit is marked in the triangle “abc.”

It must be emphasized that the main criterion is not the price of solar collectors and their installation, but also the price of a heat unit that is produced by solar collector and effectively consumed. It is possible and even necessary to calculate how much the produced heat for the individual-family house or apartment buildings will cost with the chosen type of solar collector. The application of this method allows the estimation of all technical aspects of economic projections to the levelized cost of the heat unit. Initial calculations show that the increasing heat prices make the use of solar collectors for domestic hot water preparation competitive in some cities.

Moreover, if ecological, economic and social benefit would be comprehensively evaluated in a long-term period, and on that basis, consumers that produce energy using solar energy would be given support, the demand for advanced solar technologies would increase noticeably. The increase of biofuel prices and the unification of prices in all Baltic and Northern countries indicate that this type of fuel is transported at a longer distance; the transportation costs form an increasingly smaller part of the final cost, when the prices of solid fuel are increasing. In Lithuanian market, the prices of wood fuel are similar to the prices in neighbouring states, and biofuel becomes the product that certainly cannot determine the choice, although, it is necessary to be on a way of objective knowledge and education but not on the presentation of one side of the information. It is very important in the initial period, when the first failures of inadequate decisions can block the way to advanced technologies, usage of which depends on successful examples.

The background of the research lies in the following conceptual statements:

- The support of state and private market participants to RES business is considered as a correction of market shortcomings to solve sustainable development problems;
- Prices of all energy resources must be based on social costs with the assumption that it is methodologically possible to justify all aspects in the use of fuel and energy – security, environmental, social, and renewability;
- RES benefit can be assessed according to real benefit now and in the future, which is not fixed in investment efficiency with the assumption that various sustainability criteria may have a single denominator on single methodological background;
- Justification of RES support measures is based on the comparison of fuel and energy resources used by energy system, on competitive ability while assessing various sustainable development aspects by a single criterion;
- RES competitive ability is evaluated not in a general approach, but in a specific territorial environment till a certain optimal level, where marginal costs and marginal benefits of all types of energy become equal.

With regard to the definition of the guidelines for sustainable energy development in economic understanding, firstly, current or necessary infrastructure, related to the used type of fuel should be clearly evaluated.

1.2.4. Methodology for Assessment of Energy Resource Renewability

A dynamic energy resource allocation criterion for energy opens up new prospects for the development of the theory, allowing solution to substantial and radical economic problems (Griffin and Steele, 1986). The essential difference between dynamic and static criteria is determined by immanent quality of the renewable and fossil energy resources, utilization of which could cause economic implications. Exploitation of depletable energy resources (such as oilfield) in one period affects the way it could be operated in any oncoming period. Available amount of depletable energy resources is finite; therefore, the owner is forced to deal with the question of when and to what extent to extract oil. In this way, one period is not identical to another. This question can be extended to the global, universal dimension: current utilization of depletable energy resources limits its future use. Consequently, the owner of the oil field and the user of oil and, finally, general public have to try to optimize the production and consumption in the face of dealing with the dilemma of how to economically influence these aspects. The dominant forms of energy-oil, natural gas, coal and uranium are depletable; so the concept of dynamic criteria in terms of economic theory can help show benefits of renewables.

Discounting plays a vital role in the economic policy. Definition of discount rate value can have multiple meanings. Depletion of non-renewable energy resources and its impact on environmental degradation in the future should be discounted. For this reasons, the idea to apply to RES a lower discount rate than that in the market was introduced. For example, Tahvonen and Kuluvainen (1993) demonstrated that the sustainable trajectory of sustainable energy could be achieved only in the case of a relatively low discount rate. Other scientists argue that more than 5 per cent discount rate reflects the unsustainable alternatives.

In other words, argument goes for a much lower (or even zero level) discount rate for the environment friendly technologies. This way takes into account the need for economic justice for future generations.

In summary, the results of various calculations suggest that the discount rate is usually chosen by the business based on certain measures of conjectural reasons.

Lowest rate is the banks interest rate, i.e., the so-called risk-free investment (e.g., now about 4 per cent). Adding the risk factors and the rate of inflation makes it rarely possible to expect that the discount rate calculation applied will be below 7 per cent.

An important problem in the assessment of natural resources is the utilization of correct discounting procedures. The starting point of debates is the discount rate, which is a part of the price of the system. Priority in time reflects the fact that people evaluating future events typically use discount rates. Events in the future are given a diminished weight compared to events that are happening at present. Borrowed money (financial capital) interest rate is a certain minimum level of the discount rate. The risk of possible changes or interruption in energy supply chain, and even more depletion on natural resources naturally have a higher discount rate.

"Discounting plays a key role in the economic policy. Consequently, there is long-term (started in the third decade) ongoing debate, what discount rate should be chosen. Taking into account necessity to include the depletion of non-renewable natural resources and environmental degradation in the future, the discount rate should be lower than the market" (Tahvonen and Kuuluvainen, 1993).

However, it should be noted that although the rate of depletion of non-renewable resources depends on the choice of the discount rate, this relationship is far from unambiguous. Many theoretical models show that higher discount facilitates faster depletion of resources. This hypothesis can be attributed to ideas presented in Hotteling's (1931) works. Similar conclusions about the relationship between the discount rate and the quality of the environment in the broader social context are presented by Krautkraemer (1988). He showed that a lower discount rate facilitates protection of the environment, but it also promotes the accumulation of the capital and indirectly increases the demand for natural resources. Lee and Orr (1975) have also contributed to this research. They investigated value of selection of the private discount rate for optimal effect of exploitation of non-renewable natural resources in the case of monopoly.

Renewable energy sources are a particular phenomenon: the use of the promotion is definitely one of the most important economic tasks. For example, inexhaustible and never-ending solar energy resources would be an ideal way for any society. In this case, social discount rate can be considered to be zero. However, the bank lends funds to the investor with positive interest rate, which can be considered a minimum discount rate. This problem may be solved by investment subsidies, for example, using the SF funds.

Pearce and Markandya (1989) state that "there are no simple rules to determine the socially optimal discount rate. In fact, the choice of the discount rate remains one of complicated problems in economic theory, because it brings up issues which go beyond the boundaries of the single scientific discipline."

It is therefore necessary to find other ways to solve conflict between integrity maximizing the well-being and well-being of the society in the future, i.e., the distribution of wealth between different generations. Economic theory has shown that the distribution over time can be made (at least in theory) by selection of appropriate discount rate (Beckerman, 1994). As it is seen from the data in the Table 1.1, discount rate (d) is one of the most important factors of energy cost and its price. It shows return on investment, which is required seeking to attract investors (Levitt et al., 2011). The results of research show that discount rate is different for various RES-E technologies (see Table 1.1); it is as well higher than the discount rate applied when assessing traditional energy production technologies. Differences in discount rates appear due to the fact that investors differently understand risk related to technologies (Branker et al., 2011; Hernández–Moro & Martínez–Duart, 2013).

Table 1.1. Comparison of discount rates **applied in assessments of RES technologies**

Wind technologies		Solar technologies	
Author	Discount rate, %	Author	Discount rate, %
Jaramillo et al. (2004)	7-10% (on-shore wind)	Branker et al. (2011)	0; 4.5 and 10%
Acker et al. (2007)	5.5%	Reichelstein & Yorston (2013)	7.5-8.0%
Lu et al. (2011)	7% (on-shore wind)	Wang et al. (2011)	8.14%
Levitt et al. (2011)	10% (offshore wind)	Ramadhan & Naseeb (2011)	5-10%
Gass et al. (2013)	7% (on-shore wind)	Hernández-Moro & Martínez-Duart (2013)	10%
		IEA (2010)	10-12%
		Blum et al. (2013)	12.5%
		Swift (2013)	3.34-31.60%

From information provided in Table 1.1, it is seen that discount rate applied when assessing solar PV technologies is 10-12%; however, cases when discount rate (internal rate of return, IRR) exceeded 30% were found. Discount rate for wind technologies is 7-10%. This shows that investors' perceived risk of wind technologies is lower than the risk of solar technologies.

Two discount rates are segregated in the scientific literature. These are nominal and real discount rates. The interrelationship between these discount rates is shown by Eq. (1):

$$r = \frac{1+d}{1+IR} \tag{1}$$

Here: r – real discount rate, %;
IR – inflation rate, %.

There are a lot of factors influencing discount rate. They are return rate of the investor, risk premium, time horizon, interest rate, income and asset taxes. Discount rate differs depending on the country, industry and company.

In scientific literature, several indicators and methods on how discount rate is computed can be found. When assessing public investment, it is recommended to use interest rates of short- and long-term securities issued by the country or their averages during a certain time period. In cases industry or service sector are analysed, experts recommend to apply a discount rate, which is equal to alternative capital cost. Alternative capital cost is the rate of investment returns of the best alternative. Another indicator, which can be used as a discount rate, is capital cost. Weighted average cost of capital (WACC) is a common indicator used to assess projects in energy sector. It is computed under Eq. (2):

$$WACC = r_{ec} \cdot [C_e : (C_e + P_s + D)] + r_{ep} \cdot [P_s : (C_e + P_s + D)] + (1 - T) \cdot r_d \cdot [D : (C_e + P_s + D)] \tag{2}$$

Here: WACC – weighted average cost of capital;
r_{ec} – rate of return of common stock;
C_e – value of common stocks;
P_e – value of proffered stock;
r_{ep} – rate of return of proffered stock;
D – value of loan;
r_d – interest rate of loan;
T – profit tax.

The analyses of assumptions for increasing solar energy demand have been provided. The main focus is to investigate what conditions are needed for thermal energy produced by users on the basis of solar technologies to compete with other energy sources. Conclusions have been made concerning on what scale and under what assumptions solar technologies may be competitive. The external utility may be the background for additional financial sources.

1.3. RES Impact on Various Aspects of the Country's Competitiveness in Macroeconomic Dimension

The scientific issue of the relationship between RES and macro economy is not new. Different scientists analysed various aspects of the issue. Based on the results of scientific literature analysis, three main streams of literature could be distinguished. They introduce macroeconomic indicators influencing on and impacted by the RES sector development.

The first stream of literature analyses the impact of macro economy on development of RES sector. Scientists (Ting-Huan Changa et al., 2009; Sadorsky, 2009a; Sadorsky, 2009b) investigate macroeconomic factors influencing development of RES sector and answering the question of economic conditions necessary to achieve or that are relevant seeking to increase volume of RES in the country.

Research findings of Ting-Huan Changa et al. (2009) showed that a positive causal relationship between consumer price index of energy and contribution of renewables to energy supply existed in OECD countries, where economic growth in the previous period was higher than the threshold effect value of 4.13%. This finding suggested that countries developing at faster than 4.13% rate a year could have economic surplus and allocate it towards development of RES. Thus they could achieve substitution effects towards consumption of RES in order to avoid the negative relationship between oil prices and gross domestic product (further in the text GDP).

Sadorsky (2009a) analysed 18 emerging world economies. He set that an increase in real income per habitant had a positive impact on RES volume. In the long term, when real GDP (per habitant) increased by 1%, RES consumption volume (per habitant) would increase by 3.39-3.45%. The research results revealed that there was an inverse relationship between electricity price for consumer and RES consumption volume. Electricity price for consumer elasticity of RES per habitant varied in a range of (-0.16) to (-0.70). It is worth noting the results of Granger causality test. They revealed that in a short run, electricity price changes had no impact on changes of RES consumption volume, and development of the latter factor did not make an impact on development of real income per habitant.

During the same year, Sadorsky (2009b) performed an assessment of interrelationship between RES and real income, CO_2 emissions and oil prices in G7 countries (Canada, France, Germany, Italy, Japan, Great Britain, and

USA) during 1980-2005. After two econometric time series methods – fully modified ordinary least squares (FMOLS) and parametric dynamic ordinary least squares – were applied, the author computed the long-run income, CO_2 emissions and oil prices elasticity of RES consumption volume. Independently on which method was applied, results were similar. It was found that income elasticity of RES consumption volume was relatively elastic. The highest elasticity was in G. Britain (27.54), and the lowest in Canada (3.43). Income elasticity of RES consumption volume was 8.44 in G7 countries. CO_2 emission elasticity of RES consumption volume was positive (5.23), but oil price elasticity of RES consumption volume was negative (-0.65) and statistically insignificant in G7 countries. Considering results of the research conducted by the author, a conclusion could be made that RES development in future will depend on economic growth rates and CO_2 emission volume.

The second literature stream tries to explain channels through which the impact of RES sector development passes the economy. Different scientists, who investigated this issue, mainly determined the impact of increasing RES volume on economic growth (increased growth rates of gross domestic product, GDP) or on single components of GDP; as well on employment and trade balance. To this stream of literature T. Chien & J. L. Hu (2007, 2008), J. Domac et al. (2005), A. M. Bassi et al. (2009), J. Blazejczak et al. (2011) and others scientists' works could be prescribed.

Chien & Hu (2007) showed a positive relationship between consumption of RES and improving macroeconomic indicators. In their later works, authors (Chien & Hu, 2008) set that development of RES mostly impacted on capital formation. This conclusion was drawn after the analysis of 116 economies was done by applying structured equation model (SEM). Domac et al. (2005) made an assumption that increasing utilization of RES promoted creation and development of local business, created new jobs (both in RES and related sectors) and replaced imported fuel, and as a result, trade balance improved. Bassi et al. (2009) noticed that investment in RES could have a negative impact on macroeconomic indicators: when the share of RES increased in balance, possibly, income in traditional energy sector reduced; employment as well reduced in this sector. Besides, when green energy was supported with increased feed-in tariffs, market was distorted, and total business expenditure for energy increased. As a result, competitiveness and profitability of all business exercised in the country reduced, and GDP was influenced negatively. Blazejczak et al. (2011) contradicted to this opinion. After an analysis of RES sector development in Germany and related markets was performed, they set that in a long-term (by 2030), development of RES will positively influence the economy of Germany (its employment, development of local businesses and GDP). Economic branches related to traditional energy will be transformed, and losses will be less than benefit received due to RES development.

Magnani & Vaona (2013) analysed data of GDP and energy produced from RES in separate regions in Italy during 1997-2007. They applied dynamic ordinary least squares (DOLS) method and investigated relationships between GDP, capital and labour. Research results showed that in a long run, all investigated factors contributed to economic growth of regions. Employment had the greatest impact on GDP in the regions. Employment elasticity of GDP was 0.76. Capital elasticity of GDP was 0.26. This showed that when amount of capital formation increased by 1%, GDP in the regions increased by 0.26%. RES elasticity of GDP was 0.02. This allowed making a conclusion that when RES consumption volume increased by 1%, GDP in the regions would increase by 0.02%. The importance of RES was interpreted as follows: utilization of RES is valuable, since it reduces exposure of regions against volatility of fossil fuel prices; utilization of RES as well contributes to reduction of negative impact on health and environment.

Tugcu et al. (2012) analysed data of G7 countries. They noticed that both traditional energy sources and RES had impact on economic growth of economically developed countries. Seeking to perform analysis, a production function was made. Here, income depended on physical capital and labour, as well as on research and development and fuel and energy consumed.

This problem was also investigated in Lithuania (Bobinaite et al., 2011). Authors concluded that correlation existed between RES and economic growth in Lithuania; however, RES sector due to its modesty and immaturity did not have statistically significant impact on macro economy.The third stream of literature shows that a wider consumption of RES was both impacted by various macroeconomic indicators and made an influence on these indicators. Works by Fang (2011), Yildirim et al. (2012), Apergis & Payne (2010a; 2010b; 2011) could be prescribed to this group of research.

Yildirim et al. (2012) employed Toda–Yamamoto procedure and investigated relationships between GDP and RES and its types such as energy from biomass, wood and waste, geothermal and hydro energy in USA during 1949-2010. The research results showed that statistically significant relationships existed between GDP and energy from waste. This allowed concluding that although energy from waste made only 6% in RES structure, namely this type of energy should be used when solving problems both related to waste accumulation in dumps and economic growth.

Fang (2011) analysed China's data. In the work, RES and share of RES in the fuel and energy consumption structure was reflected as one of the most significant factors having positive influence on economic wealth of nation and individuals. Capital (gross capital formation indicator was selected) and technological progress (expenditure to R&D was selected as an indicator) were other factors having a positive influence. The impact of labour (number of people employed was selected as an indicator) as production factor on GDP was statistically insignificant. Fang stated that surplus of labour force and low labour productivity were the main reasons for this. Sustaining to the idea of Kobo-Douglas production function, and applying OLS method, it was set that when RES increase by 1%, real GDP increases by 0.12%, GDP per habitant by 0.162%, annual income following per each habitant living in village by 0.444% and annual income following per each habitant living in city by 0.368%. As it is seen, a wider use of RES has a positive impact on both nation's and person's economic wealth. Calculated indicators of RES elasticity of income showed that RES development in villages could increase individual's income more than RES development in cities. Therefore, when solving issues of provision the habitants living in villages with income, it is purposeful to develop RES namely in villages.

Apergis and Payne (2010a; 2010b; 2011) investigated data of 6 Central America countries during 1980-2006, 20 OECD countries during 1985-2005 and 13 European and Asian countries during 1992-2007. They set that rising economy of these countries (i.e., increasing GDP and amounts of capital formation) influenced RES development in these countries in a short run. This allowed making a conclusion that opportunities for a broader use of RES appeared and increased when economy grew. However, authors did not set what should the level of GDP or basic capital formation be, which could influence RES development at a more rapid rate. Scientists set that RES development was also dependent on labour force in 20 OECD countries. However, the way this impact was realized was not analysed in the paper. Scientists noticed that the link between GDP, gross basic capital formation and RES was bilateral. This allowed making a conclusion that in a short run, wider RES utilization impacted GDP through its component, named "gross basic capital formation." Analysis of casual relationship in a short run was exercised by applying Granger causality test. In a long run, RES had an impact on economic growth. After FMOLS method was employed, and RES elasticity of GDP was computed, it was found that when RES consumption volume increased by 1% in Eurasian countries (Russia was not considered), economies of countries grew by 0.074%. When data of Russia were considered, RES elasticity of GDP increased to 0.195%. The results of the analysis of Central America and OECD countries revealed that RES elasticity of GDP was 0.244% and 0.76%, respectively. Research results revealed that RES sector is an important constituent part of the economies of these countries, since it contributes to a rise of economic wealth (it is measured by GDP indicator) of the nation.

In later years Apergis & Payne (2012) analysed 80 economically well-developed and developing countries by including into the model consumption volume of traditional energy sources. Research results revealed that during 1990-2007, both traditional energy sources and RES had a positive impact on GDP, and changes of GDP influenced fuel and energy consumption changes. When consumption volume of fuel and energy used in production and supply of various goods and services increased, GDP of these countries increased. Computed traditional energy sources and RES elasticity of GDP were very similar 0.384% and 0.371%, respectively. This showed that independently of the type of fuel and energy, the impact of consumption of natural resources on GDP was similar; therefore, the role of traditional energy sources is important when building economic wealth. However, the increase of consumption of traditional energy sources makes impact on climate change. Seeking to stabilize the increase of GHG emissions, a greater attention should be paid to a broader use of RES. When the analysis of causal relationships was carried out, scientists set that during the analysed time period, traditional energy sources were being substituted by RES. During the time of the research, other relations were set. It was determined that the impact of RES consumption volume on gross basic capital formation was statistically insignificant, but the impact of traditional energy sources was significant. This showed that the impact of RES on economic growth was passed through other channels.

Al-Mulali et al. (2013) analysed 108 countries. These are high-income countries (32), higher than average income countries (26) and lower than average and low-income countries (50). After FMOLS method was applied, it was set that bilateral relationship between RES consumption volume and GDP existed in 79% of countries. This is especially typical for high-income countries. On the one hand, this showed that economies of these countries highly depended on volume of energy consumed. On the other hand, countries cover energy demand by using renewable fuel and energy, which satisfies the criteria of sustainable development. Level of GDP was a very important factor in accelerating RES development.

Menegaki (2011) analysed EU-27 countries. The author computed that in a long run, RES and labour (employment was chosen as a proxy) made a direct impact on GDP in the region. The scale of impact of RES and employment was similar. When RES consumption increased by 1%, GDP increased by 4.4%, and when employment increased by 1%, GDP increased by 4.9%.

Bildirici (2013) investigated 10 developing and emerging economies during 1980-2009. After autoregressive distributed lag bounds testing (ARDL) method was employed, it was set that utilization of biomass contributed to economic growth in these countries both in a long and short run. Besides, in most cases, GDP level was important seeking to accelerate the development of biomass sector.

Thus summarizing, it can be stated that changes of RES consumption have links with macroeconomic indicators. On the one hand, macroeconomic indicators influence RES consumption volume. On the other hand, changes of RES consumption volume impact macroeconomic indicators.

1.4. Methodology for Evaluation of the Impact of RES Sector Development on Economic Indicators

Statistical methods based on the historical data of the country (or selected sectors) could be employed for assessing the development of RES and their impact on the macroeconomic indicators. This common trend could not provide a deep insight on the actual impact on these parameters by any single project. Following general trends can leave benefits of the individual project invisible to the country's macroeconomic indicators. They do not provide any tools or techniques enabling optimization of RES sector development (or introducing support measures) for maximizing the macroeconomic benefits of choosing optimal support measures, which could provide maximum macroeconomic benefit with minimum resources. Such assessment is inseparable from the microeconomic analysis of the project chosen by the individual investor (taking into account the economic benefits for the developer), utilizing the external parameters, affecting the macroeconomic indicators. Results obtained in a single project analysis can be interpolated to assess possible impact of the support scheme (country level).

However, it should be noted that all the above-mentioned methodological principles are not directly related to the fundamental problem, which is the economic competitiveness of the country. This means that income earned in the State is evaluated by GDP ratio according to international practice, but the dynamics of the asset also should be taken into account. If the ratio is achieved by a huge public debt, if the national asset, an important component of which is the forest, is decreasing, if air pollution is increasing, the quality of life is decreasing at the same time. However, very large investments are required to implement these objectives. Bearing in mind that large energy projects, such as electricity links, are exercised presently, it is essential that obligations were linked with financing abilities and economic-financial capital receiving mechanisms. The fact that it is needed to justify the scale of RES implementation mandatorily through the Directives shows that market mechanisms perform insufficiently due to undisclosed reasons. The task is to investigate these reasons. Besides, RES development cannot be treated separately from economic development tendencies. Obligations provided in EU Directives must be related to positive effect, which is the base for motivation to exercise RES programmes in many countries. Therefore, it is required to investigate the positive macroeconomic effect and various conditions, which allow assessing and linking this effect to applied support measures.

Not only directly related expenditures, but also the resulting effect must be evaluated (see Table 1.2).

It can be stated that the assessment is rather complicated, and it is not necessary to limit oneself only to the assessment of the impact on GDP.

Another benefit of using microeconomic model with external macroeconomic parameters is that it facilitates the assessment of how selected RES projects replace traditional energy resources and sectors/energy carriers they could affect. One of the drawbacks of this approach is that it is hard to assess the technological innovations and business development effect: using data from a single project, it is difficult to make assumptions about how the development of similar projects could affect the sector in the long term. For example, in the context of a wind farm or geothermal energy project, it is difficult to estimate how local business reacts if a number of similar projects reach hundreds or thousands. When looking at the first project, usually, the major part of equipment is imported, but if the number of similar projects increases, and this process gains large scope, it is likely that some of the equipment would be produced locally. This potential positive effect of continuing support could be missed by the model based on microeconomic indicators. On the other hand, experience gained in Lithuania and foreign countries could help make and aggregate estimates of possible cumulative effect, which could be expected in case of RES utilization in the country.

Table 1.2. The context of the impact of RES on macroeconomic indicators: three methods for GDP accounting in 2013 based on Lithuanian statistical data (at current prices, 2013), million EUR

	Production method (value added)	**Expenditure method**	**Income method**
Money flow	GDP = demand of final goods = value added = wage + capital income (Accounting is done agreeably to branches of the economy)	GDP = Household expenditure (C) + Investment (domestic and foreign capital) (I) + Governmental expenditure (G) + Export of goods (X) – Import of goods (Z)	GDP = Wages and Social security (W) + Income from rent of land or other real estate (R)+ Interest income (I) + Profit (P) + Taxes indirect (T_i) – Subsidies (T_s)
GDP	**34955.6**	**34955.6**	**34955.6**
Inventory Forest Expectations of people Air Water resources		Current account (balance) Net international investment position	Capital (real financial and intellectual) Savings and worth
		Budget Balance of Governmental income – expenditure ↓ Public (government) debt	

Generally, there are two appraisals (financial and economic) that consider money (other appraisals may consider safety, environmental impact, etc.). The financial appraisal tries to see if the project will work as a business proposition. In other words it considers only those costs and benefits, which are relevant to the business entity undertaking the project. A financial appraisal looks only at the effect of the project on the "bottom line" of the firm or enterprise. The economic appraisal attempts to decide whether the flow of benefits is greater than the flow of costs, measured according to broader socio-economic criteria. Factors such as employment effects, environmental impacts, balance of payment effects and poverty alleviation might be taken into account. It is possible for a project to pass an economic evaluation, but to be untenable as a business proposition; if the Government wants the project to go ahead and is ready to pay, it may decide to use some public money or "soft" finance from other sources. One solution to the problem of making projects work as business propositions has been for lenders to require "sovereign guarantees" and then rely only on economic appraisal to justify the project to the borrower government.

The World Bank almost exclusively lends in this way, and so, from their perspective, economic evaluation is the most important factor in deciding whether a project goes ahead. It is also possible for a project to fail an economic evaluation, but nevertheless to represent a good business proposition. A typical example of this situation would be where society bears the cost of negative environmental effects of a project. One important purpose of the Government regulatory or fiscal measures is to prevent a situation arising where it is profitable for a firm or a private individual to act in a way which results in a net cost to society. The externality is said to occur whenever a third party (i.e., somebody not directly involved in a transaction) incurs a cost or enjoys a benefit as a result of a transaction. The classic example of an externality is pollution; car exhaust fumes do not distinguish between drivers and non-drivers, but enter everybody's lungs indiscriminately. Car drivers reap the benefits of using petrol, but part of the cost is borne by those whose quality of life is affected by pollution. Unless specific steps are taken by the Governments to internalise these costs, for example, by introducing pollution charges, they remain external to the market. Where an energy project is sponsored by the Government, it is common for externalities to be valued as far as possible and incorporated into the project appraisal. This is because the Government should be concerned with the wider welfare implications of its activities, beyond their immediate income-generating potential. It is generally assumed that the private sector need not concern themselves with including externalities in investment appraisals, as they are concerned only with the "bottom line." However, there are two reasons why an energy project sponsor in the private sector might be concerned with externalities:

- if "soft" finance is to be sought from bodies such as the Global Environmental Facility (GEF) or bilateral aid programmes, it is very often a condition that projects should demonstrate a wider economic benefit, and externalities (either generating positive externalities or minimising negative ones) are one aspect of this;
- over the lifetime of the project, there is a likelihood that the Government of the host country might introduce or tighten regulatory or fiscal measures to control or internalise externalities, especially environmental ones. This eventuality might have a significant effect on the viability of the project.

A number of techniques are available to incorporate externalities into the investment appraisal process. The choice of technique depends to a large extent on the nature of the externality itself.

Levelized cost of energy (further in the text LCOE) is one of the most widely used approaches for comparison of different energy generation alternatives. It is based on the principle that present value of total life-cycle cost is calculated considering the chosen discount rate, and then it is allocated per one unit of energy (Bobinaite and Tarvydas, 2014). The LCOE approach is useful for assessing competitiveness of different energy generation options, possible gains for the project developer, and also it helps to make insights into macroeconomic effects (after external data are utilized). The calculation of LCOE is based on an extended LCOE approach, which is described by Eq. (3):

$$LCOE = \frac{\sum_{t=0}^{T} \frac{I_t + O\,\&\,M_t + F_t - PTC_t - ITC_t - ATL_t}{(1+d)^t} - \frac{RV}{(1+d)^T}}{\sum_{t=0}^{T} \frac{C_I \cdot 8760 \cdot LF \cdot (1-DR)^t}{(1+d)^t}} \qquad (3)$$

where: I_t – investment cost at time step t, EUR; $O\ \&\ M_t$ – operation and maintenance cost at time step t, EUR; F_t – fuel cost at time step t, EUR; PTC_t – subsidy for energy production, EUR; ITC_t – investment subsidy, EUR; ATL_t – income / cost for emission allowances, EUR; RV – residual value, EUR; C_I – installed capacity, kW (MW); LF – load factor, %; d – discount rate, %; t – life time, years; DR – degradation rate of technology, %.

The advantage of the extended LCOE approach is that it allows assessing the effect of different RES support schemes on energy cost; also it takes into account both environmental constrains and technological limitations.

Discount rate is a constituent part of the LCOE model. The results of literature analysis showed that applied discount rate is different for various technologies and assumptions that are taken into account. For evaluation of onshore wind technologies 7-10% discount rate is usually applied (Jaramillo et al., 2004; Acker et al., 2007; Lu et al., 2011; Gass et al., 2013).

LCOE depends on costs and their structure. Three groups of cost – investment cost, O&M cost and fuel cost – are analysed.

Investment cost for onshore wind technology consists of (IRENA, 2012):

- turbine cost, which includes converter, gearbox, generator, transformer, rotor, tower and other. This cost makes 64-84% of total capital cost;
- work cost – this is construction cost encompassing field preparation and tower foundation cost. This cost makes 16% of total capital cost;
- connection to network cost includes transformers and substations, as well connection to local distribution and transmission network. This cost makes about 11% of total capital cost;
- other capital cost encompasses construction of buildings, control systems, consultation and other cost. This cost makes 9% of total capital cost.

The investment cost is incurred during the time of wind technology implementation, and usually, it is sustained during a longer period of time (for example, in several years). Since high initial investment cost is necessary, a wind power technology implementer uses both own and borrowed financial resources. Seeking to fully assess cost, interest during construction is taken into consideration. Thus, total investment cost is calculated regarding Eq. (4):

$$I = IDC + I_{on} \tag{4}$$

where: IDC – interest during construction, EUR; I_{on} – overnight investment cost, EUR.

There are several different ways to compute IDC, but a linear approach is used in this paper. IDC is computed based on Eq. (5):

$$IDC = \sum_{j=1}^{k} I_j \cdot (1+i)^{k-j} \tag{5}$$

where: i – interest rate, %.

Two groups of O&M cost – fixed, which does not depend on generation output, and variable, which is directly influenced by generation output – are analysed. Variable cost also includes environmental taxes. Evaluation of O&M cost is computed based on Eq. (6):

$$O\&M_t = FC_t + VC_t \tag{6}$$

where: FC_t – fixed cost, EUR; VC_t – total variable cost, EUR;

$$VC_t = C_I \cdot 8760 \cdot LF \cdot (VC'_t + ET_t) \tag{7}$$

where: VC'_t – variable cost per unit of energy produced, EUR/kWh;
ET_t – environmental taxes, EUR/kWh.

Since RES technologies receive various forms and levels of support, as well as various financing instruments are used; therefore, LCOE takes this into account. Besides, the possibility to trade emission allowances is also foreseen (it is computed based on Eq. (8)), and degradation rate of RES technology is added to Eq. (3).

Price of emission allowances is calculated according to Eq. (8):

$$ATL_t = (F_t \cdot EF - ATL_{ft}) \cdot ATL_P \tag{8}$$

where: ATL_p – price of emission allowance, EUR/t; ATL_{ft} – number of free emission allowances; EF – CO_2 emission coefficient, t.

Thus, an extended LCOE approach allows assessing the effect of all currently existing RES support schemes on energy cost. It also allows comparing LCOE and evaluating the investment attractiveness for the investor together with possible effect to the costumers' and / or countries' goals.

Model generator was created for calculation of the economic benefits for RES projects (using developers perspective) and possible impacts on selected macroeconomic indicators, such as employment, foreign trade balance and budget revenues through taxes. Model generator approach was utilized to facilitate greater flexibility. Based on the data entered, model generates formulas describing cash flow time series, providing the user with maximum flexibility to change individual technical/economic parameters of selected technologies and to get the result immediately. Only a change in the estimated construction time or lifetime will require model to be rebuilt due to the changes in the length of the time series.

Proposed model is based on extended version of a universally recognized levelized cost method. It allows the evaluation of the project microeconomic criteria. Model was complimented by the macroeconomic benefits for the country as a plausible reason for the support measures.

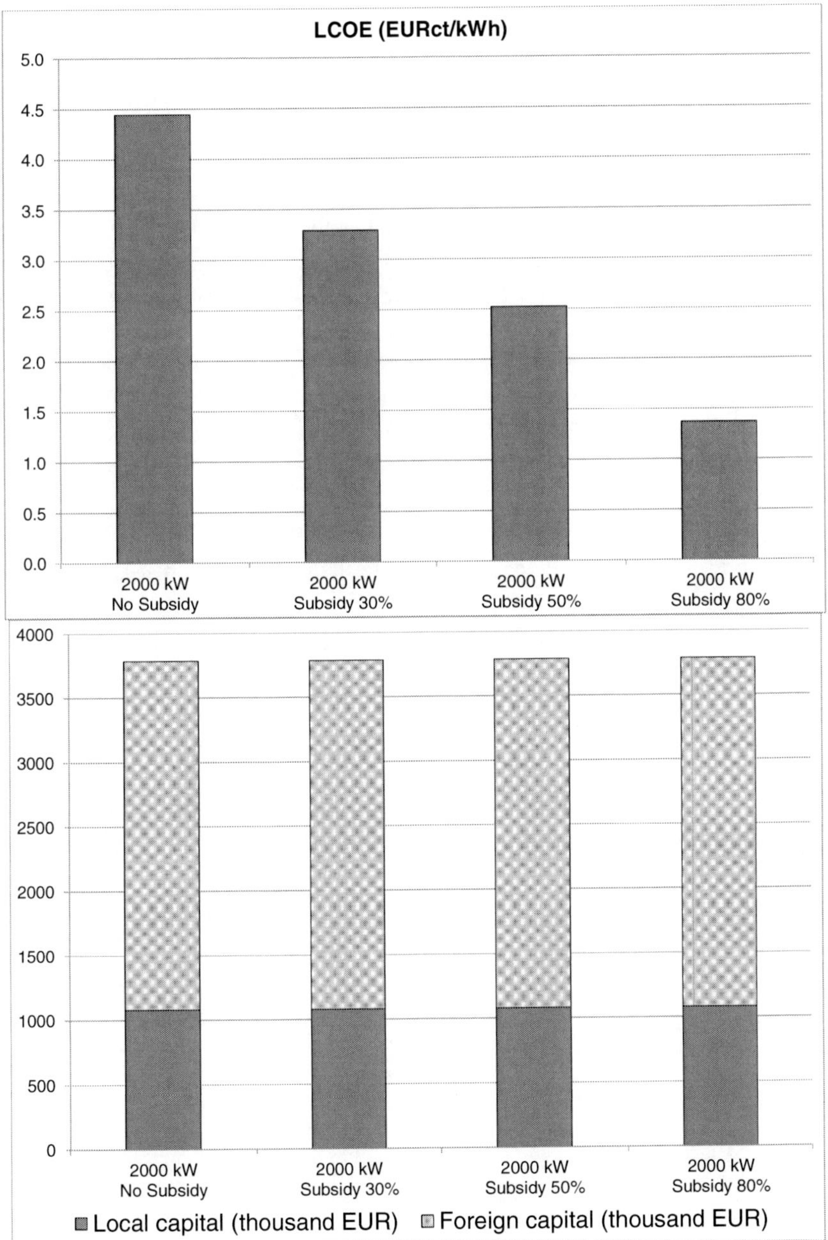

Figure 1.4. Levelized cost of energy production (LCOE) and split between imported and domestic resources.

Macro-economic benefits are estimated by means of two external parameters: the wage share and local / imported resources ratio in all the main cost components of the project. These parameters can be estimated by interviewing the project developers, based on expert judgment or external data sources.

When fuel-less technologies are analysed (for example wind turbines or solar panels), all information regarding fuel costs may be set to zero. It should be noted that some of the analysed technologies such as solar collectors, do not consume fuel directly, but use a certain amount of electricity for operation.

If the project analysed is designed to sell heat or power (to the grid or to the end-users), energy price should be specified. Market price could also be used to demonstrate investment attractiveness to self-consumption. The carbon dioxide content of the fuel is used to calculate the balance of the emission allowances; it may also be utilized for evaluating environmental impact.

The main result of the model calculation – levelized (weighted) energy cost of each project presented in the ATS sheet along with details of the project: present value of taxes paid and the balance between uses of local / imported resources (Figure 1.4).

This information helps the developer to assess the viability of projects selected under constrains provided, and support state decision-makers evaluate maximum amount of reasonable support measures and impact on the macroeconomic parameters in the long run. Projects are examined individually; therefore, users can draw a conclusion not only on direct macroeconomic impact, but also on possible impact on market shift caused by the support measures. In this way, a broader macroeconomic impact of selected technology could be evaluated.

Chapter 2

Analysis of Natural RES Influence on Macroeconomic Indicators

2.1. Projections and Dynamics of RES Consumption

After the restoration of independence in 1990, the consumption of RES was very low in Lithuania. It amounted to only 320.3 kilotons of oil equivalent (further in the text ktoe) and made only 2% of gross inland energy consumption in 1990 (see Figure 2.1).

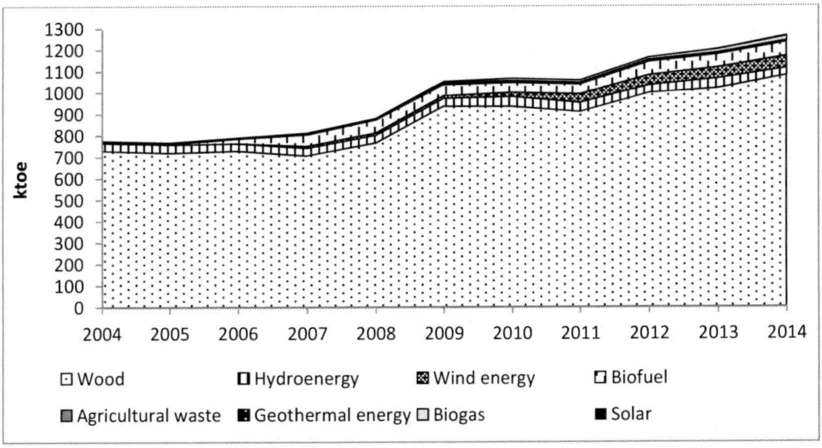

Figure 2.1. Development of RES gross inland consumption in Lithuania during 2004–2014 (Lithuanian Statistics, 2015).

Because of the implemented RES support policy, RES gross inland consumption was increasing by 6.1% a year during 2000–2012. In 2012, RES gross inland consumption amounted to 1164.8 ktoe (10.4% of gross inland energy consumption). Currently, wood and wood waste dominate in the structure of RES. In 2012, wood and wood waste covered 85.7%, biofuels – 5.4%, hydro energy – 3.1%, wind energy – 4.0%, biogas – 1.0%, agricultural waste – 0.4%, geothermal energy – 0.3% and solar energy – 0.02% of total RES consumption. 402.8 ktoe of these RES were transformed in power and heat plants for heat and electricity production in 2012.

Households are the main consumers of RES. However, during Lithuania's rapid economic development period, households were inclined to reduce RES consumption by 3.3% a year. This reduction was covered by the increased consumption of natural gas and electricity. During the period of national economy slowdown, the common structure of fuel consumed in household sector changed. Again, households increased consumption of wood and wood waste. In 2009, 71.9% of RES have been consumed in households, 11.2% – in industry, 9.2% – in transport, 5.2% – in trade and service, 1.9% – in agriculture, 0.5% – in construction. During time of recovery of economy (2011–2012), RES final consumption reduced by 1.5% in 2011 and increased by 3.9% in 2012. With reference to data of 2012, households consumed 560.9 ktoe of RES. This is 74.3% of final RES consumption. Industry consumed 10.9% of RES, transport sector – 8.1%, trade and service – 4.6%, agriculture and fishery – 1.6%, and construction – 0.5%. Increase in RES consumption was highly related to increasing investment in RES sector.

Disproportions in RES implementation are evident. Assimilation of essentially different types of energy under RES concept is one of the reasons for such disproportions. Market flaws, which do not allow mastering technologies using natural, renewable energy sources – solar, geothermal and wind – will be discussed in more detail. One more reason is that energy consumers could generate significant volumes of energy using solar or wind technologies; however, this allegedly contradicts strategic provisions of the energy sector and damages financial indicators of energy companies. Even official statistics actually nearly ignore small energy producers.

One should have in mind that technologies of various natural types of RES play different role in electricity and heat generation. Electricity generation is related to establishment of additional infrastructure. District heating systems are dominating in heat generation sector, and new biomass boiler-houses are currently under active construction. However, significant share of heat and especially hot water can be produced in solar collector systems and geothermal plants. The important factor to note is that significant use of solar and geothermal technologies is possible in households, and to be more precise, in multifamily building sector.

In 2012, National Commission for Energy Control and Prices has estimated that 4.4 TWh of supported electricity would be generated using renewable energy sources by 2020, having reached the Law-defined and supported targets of installed capacities by respective technologies, excluding supported electricity, generated by small producers (capacities below 30 kW). In case of annual 2% power consumption growth, total supported electricity volume using renewables would make no less than 40% in energy consumption structure by year 2020 (Figure 2.2).

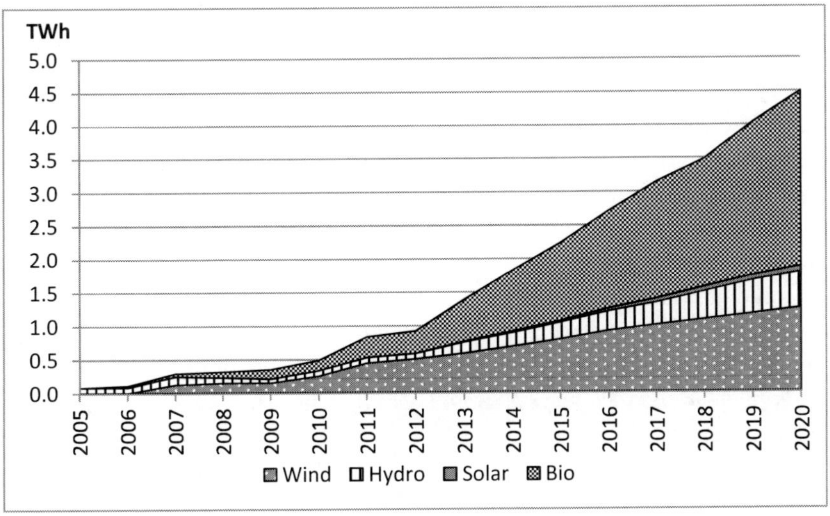

Figure 2.2. Expected dynamics of supported power, generated using RES, excluding producers up to 30 kW by year 2020, TWh.

The largest potential in power generation can be assigned to wind energy. Strategy for Development of Renewable Energy describes the volumes of electricity to be generated by 2020 using RES. The target for wind energy is 1250 GWh (500 MW) (Table 2.1). It is estimated that investment in construction of these capacities of windmills would reach 0.3-0.4 billion euro, and integration of such capacities into the national power system would increase final energy tariff by 0.26-0.32 EUR ct/kWh. Subsidy to power tariff as well as system balancing costs would influence power price.

Table 2.1. Forecasted volumes of power generation in wind power plants during period 2012-2020

WIND POWER PLANTS	2012	2013	2014	2015	2016	2017	2018	2019	2020
Installed capacity, MW	250	300	350	389	500	500	500	500	500
Electricity generation, GWh	563	688	813	924	1111	1250	1250	1250	1250

Strategy defines that on the basis of scientific research, Lithuania's economic wind energy potential would reach 200 ktoe or 2326 GWh in year 2020. This corresponds to approximately 930 MW of installed capacity.

In case total installed capacity would exceed the defined volumes by January 1, 2015, but no later, the Government is empowered to approve the Order description for further development of power plants, transmission and distribution grids, smart grids and power generation accumulation infrastructure.

Currently, SC Litgrid and Kaunas University of Technology have elaborated a common study, which defines the range of necessary infrastructure and investment till year 2030 in detail, based on several scenarios. However, all scenarios show that increase of wind energy capacities would require huge financial resources for development of infrastructure.

Implementation Plan for National Strategy on Development of Renewable Energy Sources *during years 2010-2015* defines specific measures for achieving the above targets. Measures related to development of wind energy are as follows:

- Investigate technical and economic feasibility and provide suggestions on promotion of actions to increase installed capacities of wind power plants to more than 500 MW, solar plants – more than 10 MW, biomass CHP plants – more than 162 MW;
- Elaborate legislation for planning, design, construction, connection to power grid and operation of marine wind power plants;
- Elaborate and provide suggestions concerning financial measures, which would promote installation of energy generation capacities in buildings, using renewable energy sources (biomass, solar, geothermal, hydro, wind energy);

- Elaborate and approve drafts of revised Technical Rules for connection of wind power plants to National Power System, and simplify the requirements to connect wind power plants of capacity below capacity of 250 kW.

All this is related to mastering large volumes of financial resources.

2.2. Technological Feasibility to Use Solar Energy

Inadequately small role is assigned to solar and geothermal energy potential in the country. Solar energy is a very important source of renewable energy; however, opinion dominates that economic potential for solar energy in Lithuania is restricted due to expensive technologies and inappropriate climate conditions. Such attitude has no background, as annual average solar radiation is approximately 1000 kWh/m^2 in Lithuania, while this factor is 1260 kWh/m^2 in southern Germany and 970 kWh/m^2 in northern Germany. Such data shows that climate conditions for development of solar energy in Lithuania are similar to those in Germany (Ministry of Energy of Lithuania, 2011). This is approved by the activities of companies operating in Lithuania. JSC "Arginta" (www.arginta.lt, on 28.06.2013), located in the vicinity of Riese village, has installed solar power plant with capacity of 150 kWp. The company declares that installed stationery system with capacity of 1 kW generates 880-960 kWh per year, and the next system would generate 1100-1250 kWh of electricity using sun. JSC "Rifas" (www.rifas.lt, on 30.06.2013), located in Panevezys City, has installed solar system, which is following sun with installed capacity of 7 kW and covers area of 50 m^2. One can find data on energy generated by solar system online, which shows daily or those of a chosen period electricity generating data. Performed analysis confirms that 1kW of installed capacity generates about 1173 kWh of electricity (data from 2011).

Solar energy technologies differ on the basis of purpose. The most efficient and the most frequent method is the use of sun for heat production, when system is used for preparation of hot tap water or as a heat carrier in integrated heating systems in buildings. Another use of solar energy raised huge interest in Lithuania – electricity generation using photovoltaic (PV) systems (Reichenstein and Yorston, 2013). New energy generation technologies using concentrated solar energy and direct solar heat for food production are still not used in Lithuania.

Figure 2.3 shows feasibility options to use solar energy.

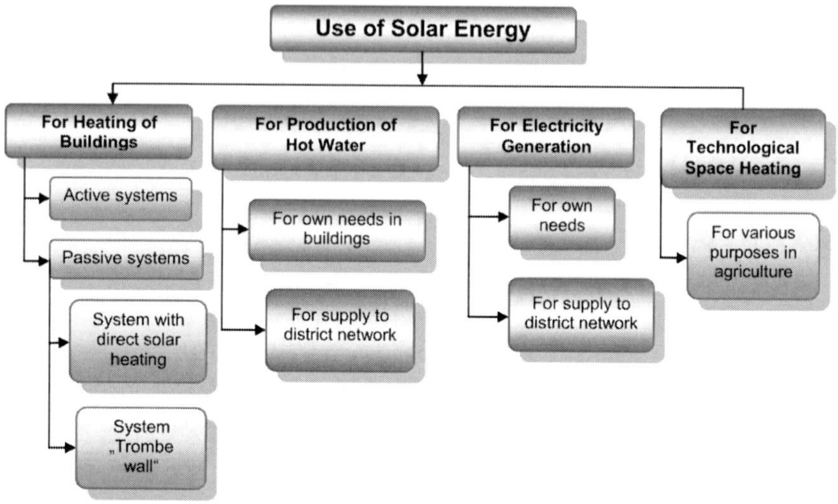

Figure 2.3. Feasibility option scheme for solar energy.

Efficiency of technologies using solar energy for heat generation reaches 85%, while the efficiency of PV systems generating electricity reaches only 10-15%. Besides, solar PV systems are more expensive, and payback period is longer. Further, several examples using solar energy will be provided, and advantages and disadvantages of the use of solar energy for heat and electricity generation will be assessed.

The volume of generated electricity per $1m^2$ of the surface of installed solar PV system is only about 165 kWh/m^2. Meanwhile the volume of heat generated by solar collectors per $1m^2$ of the surface of installed solar heat system is about 800-1200 kWh/m^2/month.

The data of Lithuanian Hydro-meteorological Service under the Ministry of Environment of Lithuanian Republic shows that long-year average full sun exposition in kWh/m^2 per horizontal surface per month and during the whole year is 939 kWh/m^2 of solar energy per year in Vilnius region. Number of sunny hours during individual months is presented in Table 2.2.

Efficiency of thermal energy generation using solar energy is 30-45%, which is much higher than for electricity generation, which is about 10-15%, thus the use of heat generation systems is much more prospective. However, the development of these systems is impeded not by investment costs, but more by lack of knowledge about technical possibilities and lack of good practice examples. While production of PV units and solar collector is becoming cheaper and the prices of natural gas and electricity generation using fossil fuel are growing, solar technologies develop faster and become more popular.

Table 2.2. Number of sunny hours in Lithuania

Location	Months												Total
	1	2	3	4	5	6	7	8	9	10	11	12	
Vilnius	41	70	126	165	243	250	243	234	150	96	42	30	1690
Nida	37	65	131	189	265	290	281	254	171	102	42	30	1857
Varena	32	64	125	167	252	260	252	243	151	92	33	20	1691
Utena	39	68	123	170	248	259	252	229	151	95	40	28	1702

As it was mentioned before, the largest share of solar energy available for conversion to heat (about 80%) falls into warmer period (March – October) and is used for preparation of hot water, heating of outside pools, partial heating of premises. Only 190 kWh/m^2 of solar energy is left for winter period. Such heating systems for buildings can be divided into passive, active and mixed (Perednis, 2000). However, such heating system must perform three main functions, which are:

- absorb and convert solar radiation into heat;
- accumulate heat, as solar radiation is not fixed and varies during the day;
- distribute heat, i.e., supply of thermal energy to heating zones, when it is required and at required volumes.

Active systems are similar to hot water preparation systems, and the difference is that the current system requires storage-accumulator (seasonal storages) to accumulate heat during summer time and use it during heating season. Huge concrete, steel reservoirs, caves, or abandoned water filled underground pits can serve as such storages. Solar heat can also be accumulated in soil using pipelines. It was estimated that 3-8 m^3 hot water storage is required for heating the area of 1 square meter. Performed investigation shows that these systems are not efficient and can be used as pilot ones only.

Passive heating systems perform all the three functions spontaneously via natural processes without forced impetus. The experience of design, construction and operation of various purpose buildings as well as international experience show that passive solar heating systems may be extremely efficient. Such systems are distinguished due to simplicity of constructional solutions and operation, economy; they do not require special staff or special equipment.

The main passive solar heating systems are as follows:

- System with direct solar heating;
- System "Trombe wall."

System with direct solar heating, the main characteristic of such system is that the absorber of solar radiation, e.g., window, showcase, is correctly directed towards the south (Perednis, 1999a; Perednis, 1999b). Usual wind glass is a good conductor of solar radiation with wavelength of 0.4 to 3μm; however, it is resistant to infrared wavelengths of bodies with inside temperature of about 10μm.

Heating system "Trombe wall" is called by the name of its creators engineer Félix Trombe and architect Jacques Michel. It is a thick wall constructed of stones, concrete and protected from outside impacts by several glass layers. Holes for air inflow and outflow are mounted near the floor and near the ceiling. The surface of the wall absorbs solar radiation; it heats and warms air between the wall and the glass. Air expands, becomes lighter and starts thermal-siphon air circulation. Warm air gets into the room via higher holes and warms it, and afterwards, cooled air returns to the space between the wall and glass, is heated again, and the cycle repeats. Due to slow natural convection and weak radiative heat flow, effective heating is felt at the distance of only several meters from the wall; thus, installation of ventilator for improvement of heat circulation is expedient.

Heating system "winter garden." Operation of this system is very similar to the "Trombe wall" system, and it also enables to maintain more stable temperature inside premises. The temperature inside the winter garden may increase significantly and fall down nearly to the outside temperature during the night. Use of solar thermal energy for space heating by means of passive installations is among the simplest and cheapest ones. While selecting appropriate passive solar heating construction, attention should be paid at selection of location of the building, direction, form, passive accumulators as well as regulation and control units. It is important to use construction materials with good thermal insulation qualities in the constructions of buildings, and after installation of passive house heating using solar energy one may state that the building will be effective and energy-efficient. Passive house heating can be installed not only in new buildings, but also in those in operation as well as outbuildings used for various purposes.

Summarizing all above, one should notify that passive heating system is as follows:

- Heating energy system for buildings, where heat access, accumulation and storage are performed in natural way using architectural units of the building with minimal volumes of additional energy and specific technical solar installations;
- System, which provides possibility to improve thermal efficiency of the building, using it together with other measures and methods to maintain good microclimate and save fuel;
- System, which does not cover full heat loading of heating season and must be used together with other heating options, using traditional energy sources;
- System, which is the most efficient in the buildings with specific design. It is also expedient to apply it to the fragments of already existing buildings, in buildings under reconstruction or newly designed ones.

With the aim to demonstrate the variety of possibilities to use solar energy solutions, here a drying chamber for medical herbs, which was installed in Panara village, Varena region in the area of Dzukija National Park, is presented. This construction is the first designed and constructed installation in Lithuania, maybe in the European Union as well. Hot air preparation costs usually make approximately 50-70 % of total costs in drying medical and spice herbs; thus, installed solar collectors significantly reduce fixed costs of the production process.

With regard to the specifics of medical herbs farm and requirements for food production technologies, special drying chamber has been designed. The building with the area of 186 m^2 has the room for initial preparation of raw plants, drying chamber, packing and packaging rooms as well as a special room for staff and a storage installed in the attic. The building also has a technical room for boiler-house and ventilators. Brick walls of building were constructed and later covered with wooden boards, and the roof of the building was covered with wooden laths.

Such system is applied in countryside areas only. Effective energy input in technological processes is estimated when average monthly solar irradiation is no less than 100 kWh/m^2, i.e., from April to September. Two types of solar collectors are used for this purpose (Genutis, 2004).

2.3. The Use of Solar Technologies in Multifamily and Individual Buildings

Solar hot water preparation systems (SHWPS) can be divided into two types. The first system has natural circulation of heat carrier, and the second system has forced circulation of heat carrier.

Recently, systems of solar collectors for preparation of hot water were most widely used in individual housing sector. This is best seen in the information available from solar panel trading companies, as well as it can be seen directly in the town districts with individual houses. However, there are not so many examples with multifamily buildings. Here, one can mention such towns as Panevezys, Alytus and Varena. What are the barriers for implementation of solar systems in multifamily buildings? It should be said that the main barriers are a lack of knowledge and understanding. A lot of faith and very little knowledge are characteristic to heat generation sector.

Decisions are made on collective basis and depend on the majority of votes; however, the majority usually has very little knowledge on the topic. The price of heat supply systems, amount of investment, pay-back period are other important issues. As long as the decision-making process is guided by non-professionals, but by the majority of residents, this process will still be sluggish. Heat and hot water bills only lead to the ideas of how to avoid payment or how to get social support. Meanwhile, people all around the world implement a lot of solutions, which would ensure renewable energy options.

2.3.1. Economic Evaluation of the Use of Solar Thermal Energy in Multifamily House

In order to estimate real possibilities and assumptions of solar energy use, for which higher demand for solar energy technologies could be expected, a specific apartment building was chosen (Klevas et al., 2012). It has five floors, 70 flats, was built in 1978 in Saules str. in Kaunas city. The decision was made to renovate it without disconnecting the DH system. There is a possibility to use the roof of the apartment building for solar collectors, and this system can provide building with hot water.

Table 2.3 shows data from personal accounts of "Kaunas energy" from December 2009 to November 2010. Current price of heat in Kaunas is 0.09 €ct/kWh including VAT.

Table 2.3. Consumption of heat in the apartment building from December 2009 to November 2010

Month	Quantity of consumed heat:			
	For space heating, kWh	For the preparation of hot water, kWh	Quantity of heat kWh/m^2	Total (useful) area, m^2
December	78150.92	21609.08	20.50	**3811.96**
January	124554.74	19945.26	32.67	
February	95428.62	1994.38	25.03	
March	77357.74	19792.26	20.29	
April	18538.20	19411.80	4.86	
May	0	17026.48	0	
June	0	19377.12	0	
July	0	18039.97	0	
August	0	18119.97	0	
September	0	23920.02	0	
October	45446.74	18823.26	11.92	
November	62092.00	18198.00	16.29	
Total:	**501568.96**	**216257.6**	**131.58**	

Project Investments

Calculations are conducted within the frame of the project MIP-2011/007. This project enabled to purchase a calculation programme. A more accurate and comprehensive information can be calculated using specialized programmes, such as Polysun, T*Sol and others. Such programmes have huge databases that allow reliably estimating basic information for technical and economical calculations. Modelling is done by evaluating the demand of hot water, working period of the flat and vacuum solar collectors in a year as well as their efficiency. Figure 2.4 shows solar energy production each month during a year. Depending on the type of the collector, its area, and heat loss parameters, the programme calculates the price of hot water heated by collector, the payback period of the system, etc.

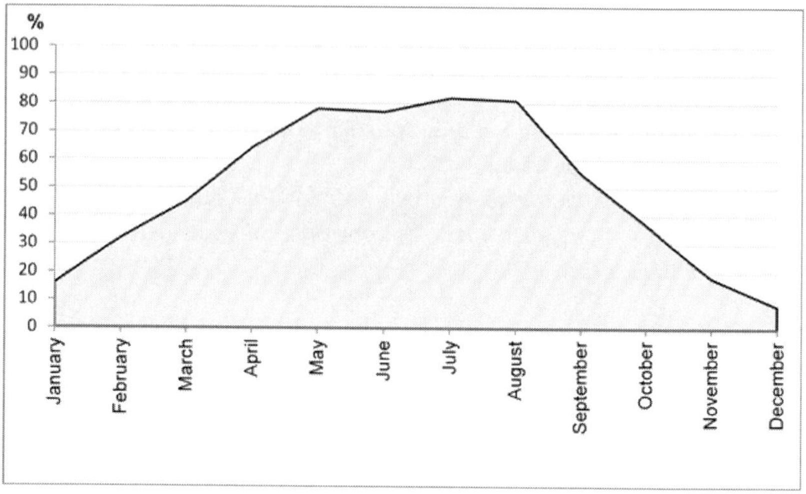

Figure 2.4. Solar energy production by month, %.

Specialised calculating programmes of solar hot water evaluate theoretical possibilities of the systems, which are often different from the real life conditions. We can state that the efficiency of the preparation systems of solar hot water depends not on the calculated quantity of thermal energy in collectors, but on the quantity of energy that is actually used. These quantities for the preparation of hot water by flat or vacuum collectors vary from 330 to 580 kWh/m^2 in Lithuania. The main advantage of the use of solar energy technologies is that exploitation is minimal. After the project implementation, expenditures of solar collector system operation and maintenance costs should be taken into consideration. For instance, to ensure planned heat production in the solar collectors, it is necessary to clean them at least once a year; expenses for hiring high-scalers can be around 290 €/year. Moreover, expenses result from changing the non-freezing heat transfer medium in system of solar collectors every five years (without changing, there is a risk of freezing). The requisite quantity for the analysed system of propylene glycol is approximately 160 litres, so the expenses would be 930 € (or 186 €/ year).

Table 2.4 presents investments of system of solar collectors for the preparation of hot water.

The main component of operational and maintenance costs is depreciation expenses. Their value depends on investments in equipment and on the lifetime of equipment. Total operational and maintenance costs were calculated in two versions: a) when a building produces 40% of heat for the preparation of hot water, the total amount is 1564 € per year and b) 50% is 2143 € per year.

Table 2.4. Investment volume for installation solar collectors for the preparation of hot water

N°	Name of equipment and materials	Hot water prepared in the systems of solar collectors (%) of total demand, €	
1	System of solar collectors for the preparation of hot water	40%	50%
Total (without VAT):		78000	107127
2	Additional materials for installation of solar collector system	3900	5356
3	Design work	2340	3214
4	Installation, adjusting work	15600	21426
Total (without VAT):		99840	137123
VAT 21%:		20966	28796
Total (VAT included):		120806	165919

Calculation of levelized costs of energy (LCoE) is carried out in these variants:

1. Residents install a collector using their own funds and loans, without a subsidy;
2. Residents get 15% subsidy for investments by the Programme for the renovation/upgrading of multifamily buildings;
3. Hypothetically, it is assumed that residents get 50% subsidy for investments by the Programme for the renovation/upgrading of multifamily buildings and environment protection programmes.

Furthermore, for calculations it is assumed that deficient quantity of heat needed for the preparation of hot water is obtained from DH network at 9 €ct/kWh. Summary of LCoE calculations is given in Table 2.5.

The calculations showed that in comparison with natural gas, 15315 m^3 of natural gas would be saved, and the emission of CO2 would be reduced by 37238 kg (about 37 tonnes of CO_2).

Table 2.5. Summary of LCoE calculations

Discount rates and level of subsidy	40% hot water €/kWh	50% hot water €/kWh
1. LCoE when discount rate is 4%		
1.1. Without subsidies	0.12	0.13
1.2. With a 15% subsidy	0.10	0.11
1.3. With a 50% subsidy	0.07	0.08
2. LCoE when discount rate is 7%		
2.1. Without a subsidy	0.14	0.16
2.2. With a 15% subsidy	0.12	0.14
2.3. With a 50% subsidy	0.08	0.09

In summary, it can be stated that with existing conditions, the use of solar energy for individual users is economically unacceptable. Nevertheless, only about 30-50% of compensations for investments could make the use of solar energy more attractive. On the other hand, it can be stated that the price is not a strictly fixed value. It is a result of many different factors and can considerably vary depending of the amount of support.

2.3.2. The Use of PV Panels for Electricity Generation to Cover Own Needs in Multifamily Residential Houses

Possibility to use electricity for own needs in residential houses appeared quite long ago (Swift, 2013). Currently, modern technologies do not only allow becoming quite independent of the national power grid for most household consumers, but allow becoming electricity suppliers as well. Nevertheless, the individual energy sector does not solve energy supply on the national scale; thus, the main load will be assigned to commercial power generators, which should not be very large. However, individual energy sector provides opportunity to a certain share of consumers to be independent of the outside grid and to gain a small direct benefit.

Since feed-in tariffs for power generated in solar plants are reducing due to cheaper technologies and lower investment costs, National Commission has defined investment volumes, required for installation of 1 kW solar power plant. These data are presented in Table 2.6.

Table 2.6. Investment volume for installation of 1 kW solar power plant

Power capacity	2013 II quarter	2013 III quarter	2013 II quarter	2013 III quarter	2013 II quarter	2013 III quarter
	Construction works		Installations		Total:	
kW	EUR		EUR		EUR	
1	396	115	241	1385	1247	1781
2	791	229	481	2770	2495	3561
3	1187	344	722	4155	3742	5342
4	1582	458	962	5540	4990	7122
5	1978	573	1203	6925	6237	8903
6	2374	687	1443	8310	7484	10684
7	2769	802	1684	9695	8731	12464
8	3165	917	1925	11080	9979	14245
10	3956	1146	2406	13850	12474	17806

As it can be seen, installation of a solar plant is still feasible. Let us investigate some practical aspects of such installation. Total electricity consumption for 3 different multifamily residential buildings has been estimated. For some of them, electricity is consumed for staircase lighting only; for others, consumption also covers operation of elevators or private natural gas boiler-house. The estimate is provided in Table 2.7.

Table 2.7. Characteristics of the buildings and total electricity consumption

	Description of the building, purpose of electricity consumption	Electricity consumption, kWh
1	Four-storey building with 32 apartments. Lighting of staircases	≅700
2	Nine-storey building with 72 apartments. Lighting of staircases + elevators	≅5700
3	Five-storey building, area 4,785 m^2. Lighting of staircases + gas boiler-house	≅11200

Data provided in Table 2.7 show that electricity demand in presented buildings varies significantly. The residents of these buildings have an opportunity to use electricity generated by solar plants.

Using software tool PV*express 3.0, power generation, the efficiency of the system, covered area (m^2) were estimated. The capacity of solar panel is 230W.

Performed estimate shows what volume of power could be generated by solar plant with installed capacity below 10 kW, and when all generated power is used to cover individual needs. In this case, energy tariffs applied to the consumers of the first group (household), which are connected to low voltage power grid, were used. In case one could use all generated electricity for private needs and without the need to pay interest for bank loans, all investment in such power plant would be returned in 9.4 years.

In other case, when 50% of electricity is used for private needs, and excess power is supplied to power grid with applied fixed feed-in tariffs, which were defined by National Commission for Energy Control and Prices and valid for the day, then excess power is supplied to the grid. In this case, the payback period would be 10.7 years, excluding state fees for income from sold electricity. Law on RES state that excess power should be purchased no longer than 12 years since the moment the producer of excess power has signed the connection to the grid contract with the grid operating body. Estimate shows that current highly favourable conditions make the plant payback period of $\cong$10 years too long. The period no longer than 8 years is acceptable for any investment.

2.4. Development of Wind Energy Technologies in Lithuania

2.4.1. Dynamics of Wind Energy Technologies in Lithuania

Development of wind energy and independent energy sector started ten years ago in Lithuania, when the first industrial range wind power plants were installed and supplied electricity to the national grid. The investigations on wind energy resources were insufficient at that period; thus, implementation of the first pilot projects had the aim to make sure whether wind was sufficient for development of wind energy in Lithuania.

Later, the order for promotion of technologies using renewable energy sources was initiated, and conditions for implementation of feasible wind energy projects gradually appeared. The indicators of wind energy stock and dynamics of energy generation are presented in Table 2.8.

Table 2.8. The development of wind power plants in Lithuania during 2004-2012

	2004	2005	2006	2007	2008	2009	2010	2011	2012
Installed capacity, MW	0.215	1	49	52	54	90	155	202	274
Power generation, GWh	N/A	1.78	13.7	106.1	131	157	202.2	472.5	537
Share in power consumption, %	N/A	0.15	0.1	0.8	1.1	1.4	1.9	4.4	5

Wind power plants have generated and supplied to the grid 537 GWh of power in 2012. This made approximately 5% of total power consumption in Lithuania. Total installed capacity of wind power plants reached about 274 MW in 2012 in Lithuania. SC Litgrid announced that total power generation from renewable energy sources was approximately 1156 GWh in 2012 in Lithuania. This made about 10.8% of total national power consumption. The input of wind energy is the largest one, about 47% (Figure 2.5).

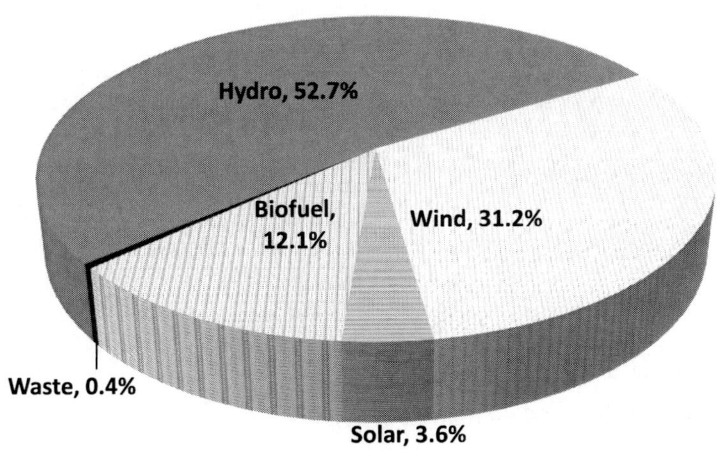

Figure 2.5. Power generation using renewable energy sources in Lithuania in 2014 (source: SC Litgrid).

On February 22, 2013, the auction on promotional quota distribution had finished for generating plants using wind energy with installed capacity over 350 kW and those connected to distribution grid. Total power generation capacity quota distributed during the auction was 150 MW (3 winners). Fixed promotional feed-in tariff of 24.5 ct/kWh will be applied for the winners from the day of power generation license. Quotas are still available for about 70 MW.

In March 2013, the auction on promotional quota distribution had finished for generating plants using wind energy with installed capacity between 30 and 350 kW and those connected to distribution grid. Total power generation capacity quota distributed during the auction was 4.83 MW (15 winners). Fixed promotional feed-in tariff of 36 ct/kWh will be applied for the winners from the day of power generation license. Quotas are still available for about 45 MW.

2.4.2. Categories of Wind Power Plants Used In Lithuania

Wind power plants by installed capacity until April 1, 2013 in Lithuania were divided into three categories (according to the Law of RES): more than 350 kW capacity, from 30 to 350 kW, and up to 30 kW capacity. From April 1, 2013, differentiation of capacity changed: wind power plants up to 10 kW capacity are considered small.

More than 350 kW capacity. Wind power plants (WPP) of this category make the majority in Lithuania. Mainly, WPP are built as a single high power unit (from 600 to 2000 kW), or as a wind farm that consists of several WPP (from 2 to 19 units). The planned wind farms are more than 100 WPP. As the world's wind energy future is associated with the development of high-power WPP, producers' supply of this category of equipment is large, and it is possible to choose from more than ten largest producers. Enercon company is a WPP market leader in Lithuania; its WPPs are an absolute majority. Today, 15 wind farms (Table 2.9) operate in Lithuania. The total capacity is 268.61 MW. 238.61 MW is connected to the transmission network (110kV); 30 MW is connected to the distribution network (35 kV and 10).

a) *From 30 to 350 kW capacity.* The number of this category of WPP has increased in recent years, because the Law of RES allows simplification of the conditions for WPP projects of this capacity. Moreover, in the beginning of 2012, NCC approved the maximum feed-in tariff 36 ct/kWh for electricity, while more than 350 kW WPP have8 €ct/kWh feed-in tariff. The total capacity of this category of WPP amounted to 12.3 MW at the end of 2012. Supply of this category of WPP is small in the market. Investors could choose from several options, such as new (Enercon E33, Vergnet GEV MP, SIVA 50/250) or used wind turbines (Bonus, WindWorld, DanWind, etc.). The choice is usually determined by the amount that could be available for the investment and investors' estimates. New WPPs cost more than twice as much as used ones; however, they produce more energy and require less maintenance; therefore, operating costs are significantly lower. Selection of the used WPP could lower initial investments, but they produce less energy (older technology, WPPs were used for 20 years), also maintenance is more expensive, and the project has a higher risk. The most popular WPP in Lithuania are 330 kW Enercon E33, which have a maximum power limited to 250 kW during operation. The reason for the limitation is the fact that permits to expand the capacity of power generation had been obtained, and EIA (Environmental Impact Assessment) procedures had been introduced before the new Law of RES and implementing legislation came into force, which changed maximum power limit from 250 kW to 350 kW. Therefore, installed capacity of the E33 WPP unit results in high relative price of about 1910 EUR/kW in comparison to prices of other producers.

b) *Up to 30 kW capacity*. WPP of this category are commonly used for the individual energy needs, because experience shows that despite the relatively high feed-in tariff for electricity (11 €ct/kWh) and free of charge connection to the distribution network, commercial projects are not economically attractive. Reasons for this are high investment per unit of power, small amount of produced energy, investment rate of return, frequent malfunctions of equipment, expensive maintenance. For example, the construction of 255 kW capacity WPP would cost approximately 0.36 million EUR (price for 1 kW is 1450 EUR, 50 kW capacity WPP about 115 thous. EUR (price for 1 kW is 2310 EUR), the price of vertical axis WPP of 5 kW capacity could be about 15-30 thous. EUR (price for 1 kW is 3-6 thous. EUR).

c) According to the SC Litgrid data, on February 2013, one WPP of 30 kW capacity (Salteniu WPP) obtained the permission to produce electricity.

Table 2.9. Wind farms and single high-power wind turbines (grey colour) in Lithuania (2012 December)

Producer	Location	MW	Model of wind power plant	Number of wind power plants	Year of the production permission
Akmeneliu WPP	Pakruojo r.	6	Enercon E82	3	2010
Anuziu WPP	Taurages r.	4	Enercon E82	2	2007
Benaiciu 1 wind farm	Kretingos r.	34	Enercon E82	17	2010
Kreivenu WPP group	Taurages r.	20	Enercon E82	10	2009
Kreivenu II WPP	Taurages r.	10	Enercon E82	5	2010
Kreivenai III	Taurages r.	15	Enercon E82	7	2011
Laukzemes WPP	Kretingos r.	16	Vestas V100	6	2006
Liepynes wind farm	Kretingos r.	9.13	Enercon E82 (E53, E33)	6	2009
Mockiu WPP	Silutes r.	12	Enercon E82	6	2010
Prysmanciu WPP	Kretingos r.	0.8	Enercon E48	1	2006
Sedos E-40 WPP	Mazeikiu r.	0.6	Enercon E40	1	2007
Sedos E-48 WPP	Mazeikiu r.	0.8	Enercon E48	1	2008
Sudenu WPP	Kretingos r.	14	Enercon E82	7	2012
Vydmantu wind farm	Kretingos r.	30	Enercon E70	15	2006
Vydmantu WPP	Kretingos r.	0.63	Enercon E40	1	2004
Seiriju wind farm	Lazdiju r.	6	Enercon E82	3	2011
Vydmantu WPP Nr. 2	Kretingos r.	2	Enercon E82	1	2011
Silales wind farm	Silales r.	13.8	Siemens	6	2012
Pakertu wind farm	Kaisiadoriu r.	6	Enercon E82	3	2012
Didsiliu wind farm	Silutes r.	21.6	Enercon E82, E48	10 + 2	2012
Ciuteliu wind farm	Silutes r.	39.1	Enercon E82	17	2013
Total amount of wind farm		**268.61**		**123**	
Total amount of single high-power wind turbines		**8.83**		**7**	

d) *Up to 10 kW capacity.* According to the procedure and conditions prescribed by the government, excess electricity produced in up to 10 kW capacity WPP is mandatory for purchasing at the fixed feed-in tariff (10 €ct/kWh). Excess electricity is recognized as no more than 50 per cent of produced electricity during the calendar year in the customer's electricity network installed WPP, which uses RES for the production of electricity. Permission is unnecessary for the expansion of such capacity electricity generation if electricity is produced only for their private use and needs of the household. Producers, which foresee the development of up to 10 kW electricity production capacity and which produce electricity only for their private use and needs of the household, must notify the electricity network operator according to the procedure of Government or its authorized institution.

e) According to the LESTO data, only a few WPP of such capacity are installed in Lithuania. The single ER-2 WPP of 2 kW capacity that started operation in April 2013 is registered in SC Litgrid database.

The Study of Analysis of Feasibility of Wind Power Plant Development indicates that WPP should not interfere with the electrical network operation and allow for the proper performance of their functions; also it should not reduce the quality of electricity and security of supply. WPP with a capacity of up to 30 kW must comply with the existing standards for this type of equipment. Moreover, they must be equipped with a separate commercial metering and fulfil the simplified technical design specifications of distribution network. All wind power plants with a capacity greater than 30 kW must meet the general requirements of the connection rules and technical design specifications of distribution network operator.

2.4.3. Wind Power Sector Development in Lithuania and Future Prospects

After the recovery of independence in 1990, the consumption of RES was very low in Lithuania. Owing to a very high dependence on import of primary energy sources from Russia, the Lithuanian Government since early 1990s was concerned with political and economic consequences of this dependence. Therefore, stimulation of increased use of RES was one of the main strategic objectives in the country's energy policy. Various estimations and strategies show that the largest potential in power generation from renewables can be assigned to wind energy (see Figure 2.6).

Estimations provided in Lithuanian National Renewable Energy Action Plan (2009) disclose that electricity produced in wind power plants (WPPs) should make 42.3% of total electricity produced from RES in 2020. In 2020, 1250 GWh of wind electricity should be produced. The latest data show that 600 GWh of wind electricity has already been produced in Lithuania in 2013, and wind electricity covered 42.1% in structure of total electricity produced from RES.

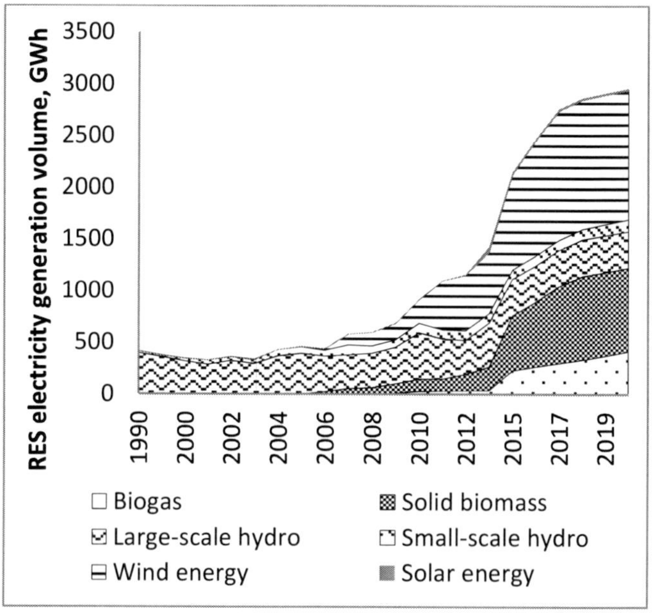

Figure 2.6. Available and projected dynamics of volume of electricity produced from RES during 1990-2020, GWh (Lithuanian National Renewable Energy Action Plan, 2009).

Strategy for Development of Renewable Energy (2010) describes the volumes of electricity to be generated by 2020 using RES. The target for wind energy is 500 MW (1250 GWh). It is estimated that investment in construction of these capacities of WPP would reach 0.29-0.41 billion EUR, and integration of such capacities into the national power system would increase final energy tariff by 0.3-0.4 EUR ct/kWh. In 2012, National Control Commission for Energy and Prices (NCCEP) has estimated that by 2020, 4.4 TWh of supported electricity would be generated using RES, after having reached the targets of installed capacities by respective technologies, excluding supported electricity from small producers (capacities below 30 kW). In case of annual 2% power consumption growth, by 2020, total supported electricity volume using renewables would make no less than 40% in energy consumption structure. Strategy defines that on the basis of scientific research, Lithuania's economic wind energy potential would reach 200 ktoe or 2326 GWh in 2020. This corresponds to approximately 930 MW of installed capacity.

The Law on Renewable Energy Sources (2011) states that by 2020, the share of electricity generated using RES should increase no less than by 20% comparing to the total final electricity consumption. For implementation of the above target, total installed capacity of WPPs should increase to 500 MW (excluding small wind turbines with capacity up to 10 kW). This limit was assumed on the basis of the study elaborated by Lithuanian Energy Institute in 2009 (Analysis of wind power development possibilities, 2009). The study states that this volume of wind power capacities can be integrated with existing reserve capacities and possibilities of transmission grid.

2.5. Evaluation of Wind Power Plants Used in Lithuania

2.5.1. Evaluation of Wind Power Plants with Installed Capacity of More Than 350 Kw

WPPs with installed capacity of more than 350 kW make the majority in Lithuania. They are mainly built as stand-alone high power units (from 600 to 2000 kW) or as wind farms consisting of several WPPs (from 2 to 19 units). Wind farms with more than 100 WPPs are in the planning stage. As the world's wind energy future is associated with the development of multi-megawatt WPPs, manufacturers' supply of this category of equipment is large, and there is a possibility to choose from more than ten largest producers. Enercon company is the leader in WPP market in Lithuania. It takes the largest share of WPP market (see Figure 2.7). These WPPs are relatively expensive, but they have proven their reliability and efficiency in moderate winds; therefore, they gained popularity in Lithuania.

Today, 15 wind farms operate in Lithuania. The total installed capacity of WPPs amounts to 284 MW. 222 MW of them are connected to the transmission network (110 kV) and 62 MW to the distribution network (35 kV, 10 kW and 0.4 kV). LCOE calculations showed (see Figure 2.8) that under the current market conditions, 2000 kW WPP could generate electricity below the market price (4.45 EUR ct/kWh versus 4.89 EUR ct/kWh in 2013).

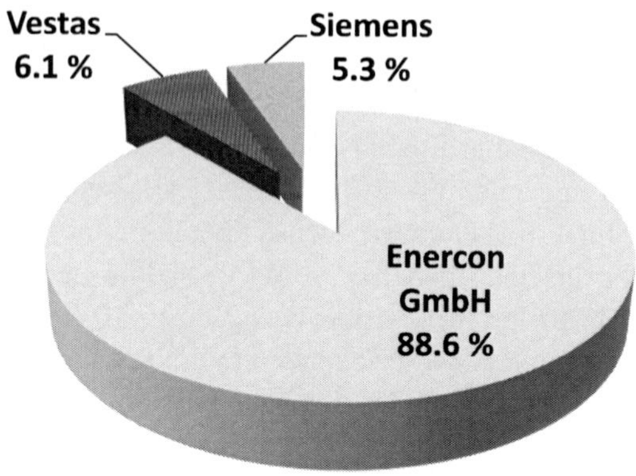

Figure 2.7. The market structure of large wind turbine suppliers in Lithuania in 2014.

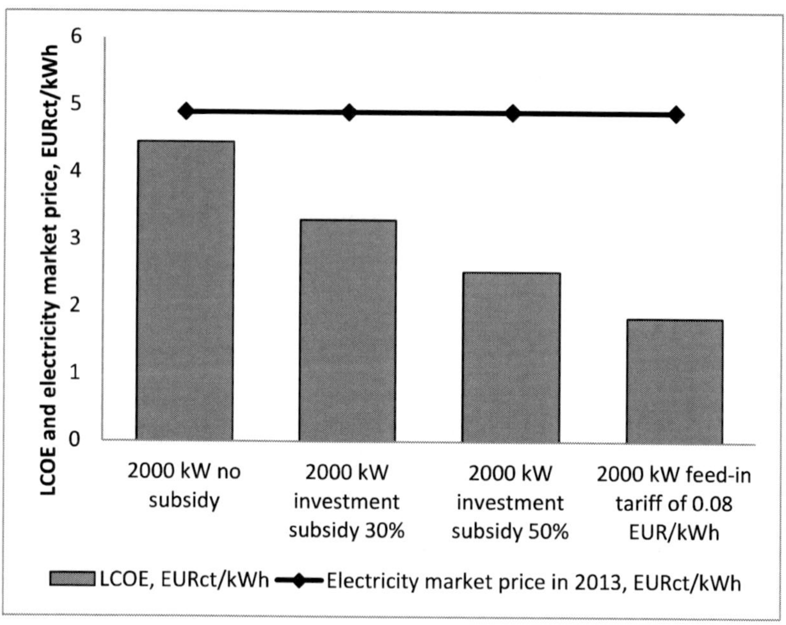

Figure 2.8. LCOE of 2000 kW onshore wind turbine subject to different support schemes and their levels (own computations).

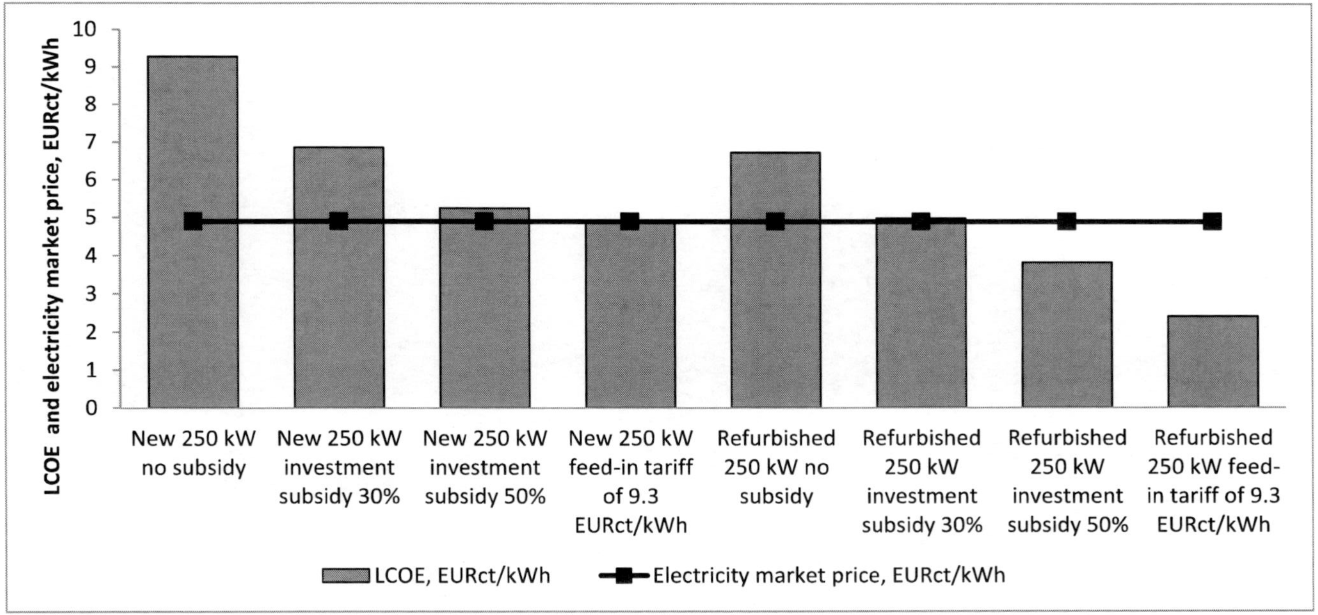

Figure 2.9. LCOE of 250 kW wind turbine (new or refurbished) subject to different support schemes and their levels (author's computations).

However, 0.44 EUR ct/kWh difference between electricity market price and LCOE is not enough to prove a business case for the private investor. Thus, support scheme is needed for this category of WPPs. Two support schemes are analysed, i.e., feed-in tariff and investment subsidy. Feed-in tariff is a currently applied support scheme for this category of WPP. Maximum feed-in tariff set by NCCEP was 7.53 EUR ct/kWh in 2013. The results of computations showed that subject to this support scheme and level, LCOE of 2000 kW WPP would be 1.89 EUR ct/kWh or by more than 2.5 times lower than electricity market price. The effect of investment subsidy is different. Subject to 30% investment subsidy the LCOE of WPP would reach 3.29 EUR ct/kWh and would be by 26% lower than electricity market price. LCOE would decrease to 2.52 EUR ct/kWh if 50% investment subsidy would be applied (see Figure 2.8).

According to available data, it was calculated that during construction or lifetime of 2000 kW wind turbine, payments to the social security fund could reach 137.2 thousand EUR. Besides, 51.2 thousand EUR would be paid to the State budget as personal income tax. It is worth noting that income/profit tax would be collected, too. Its value could vary in a range from 72.1 thousand EUR (without any subsidy) to 945.8 thousand EUR (subject to the approved feed-in tariff support scheme). Moreover, 2000 kW WPP would save 43 thousand tons of CO2 (calculations are done by taking into account the assumption that WPP would change natural gas-based PP) during its lifetime. Value in monetary terms would be 182 thousand EUR. It is worth noting that operation of 2000 kW WPP would impact on the country's trade balance. It was computed that if electricity would be produced domestically, then the requirements for imported electricity would reduce, i.e., subject to substitution of imported electricity with electricity produced in WPP, then "import position of the trade balance" could be reduced by 4.9 million EUR.

2.5.2. Evaluation of Wind Power Plants of Installed Capacity from 30 to 350 kW

The number of WPPs of installed capacity from 30 kW to 350 kW has been increasing during the recent years. The reason of this is the Law on Renewable Energy Sources (2011). It allows simplification of the conditions for WPP projects of installed capacity from 30 kW to 350 kW. Moreover, NCCEP at the beginning of 2013 approved the maximum feed-in tariff of 9.3

EUR ct/kWh for electricity produced in WPP of installed capacity from 30 kW to 350 kW. Total installed capacity of 30-350 kW WPPs amounted to 22 MW at the end of 2014. Supply of this category WPPs is small in the market. Investors usually choose from several options, such as new (Enercon E33, Vergnet GEV MP, SIVA 50/250) or second-hand wind turbines (Bonus, WindWorld, DanWind, Vestas, etc.). The choice is usually determined by the amount that could be available for the investment and investors' estimates.

While 2000 kW WPP could compete in current electricity market in Lithuania, situation with 250 kW WPP is slightly different. Computations of LCOE revealed (see Figure 2.9) that neither new nor refurbished 250 kW WPP would be competitive in electricity market if no support scheme were applied. LCOE of not supported new 250 kW WPP would be 9.3 EUR ct/kWh and of the refurbished WPP 6.7 EUR ct/kWh. These LCOE are almost twice higher than the electricity market price. Neither 30% investment subsidy nor 50% investment subsidy would be attractive to investors deciding about investment in new WPP, since LCOE of this WPP would still be higher than electricity market price. 9.3 EUR ct/kWh feed-in tariff justifies LCOE of new WPP with electricity market price. 50% investment subsidy or 9.3 EUR ct/kWh feed-in tariff would be an attractive option for the investor who thinks about construction of refurbished WPP.

Economic activity of new 250 kW WPP is not profitable in Lithuania under conditions analysed in the paper. Only in the case a refurbished 250 kW WPP were installed, such WPP could generate profit. Moreover, existing 9.3 EUR ct/kWh feed-in tariff is attractive for the WPP, since it could earn 229.1 thousand EUR during the lifetime. 50% investment subsidy could influence the achievement of 100.4 thousand EUR profit during the lifetime of WPP.

According to available data and calculation results, 250 kW wind turbines have impact on the country's trade balance. Taking into account the price of imported electricity and possibility of wind electricity to substitute it, it was computed that import position of trade balance could be reduced by 445 thousand EUR.

2.5.3. Evaluation of Wind Turbines with Installed Capacity up to 30 kW

WPPs of this category are commonly used for individual energy needs, because experience shows that despite the relatively high feed-in tariff for electricity (10.7 EUR ct/kWh in 2013) and free of charge connection to the distribution network, commercial projects are not economically attractive. Reasons for this are high investment per unit of power, small amount of produced energy, investment rate of return, frequent malfunctions of equipment and expensive maintenance.

For small wind turbines of up to 10 kW, there is a separate support scheme waiting for approval.

Wind turbines with total installed capacity of 66 kW are currently registered in the SC Litgrid database. According to the small wind power market participants, several hundreds of kW of small wind turbines are installed in Lithuania, but they are not connected to the grid.

The calculations of LCOE revealed (see Figure 2.10) that small-scale (4 kW and 10 kW) wind turbines could not compete in the current electricity market in Lithuania if investment subsidy or subsidy for electricity production were not provided. The existing feed-in tariff would be favourable for installation of 4 kW and 10 kW WPP. Furthermore, LCOE of 10 kW WPP was found to be 2.5 lower than LCOE of 4 kW when feed-in tariff was applied. In case 10 kW WPP was considered for installation, this category WPP would be attractive with 30% and more investment subsidy.

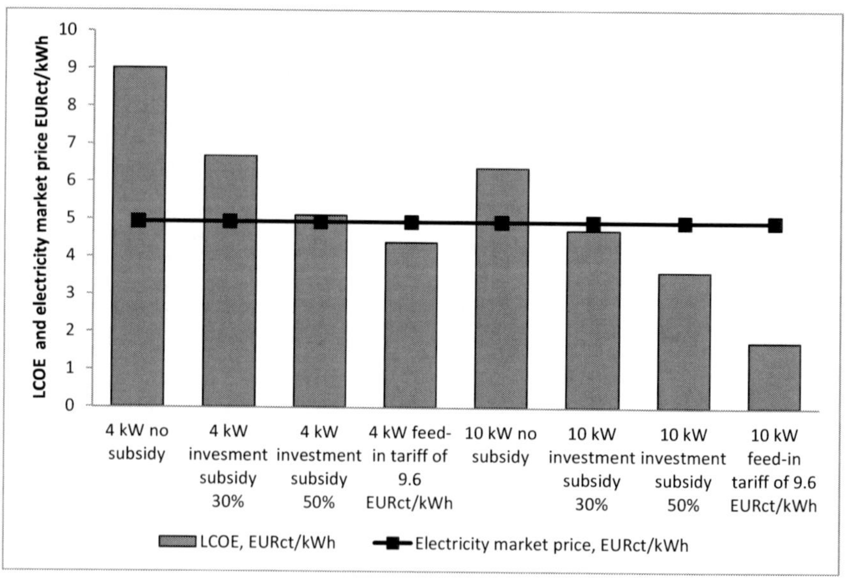

Figure 2.10. LCOE of a 4 kW and 10 kW onshore wind turbine with different level of support (author's computations).

Figure 2.10 demonstrates profits earned by the owners of 4 kW and 10 kW WPPs subject to different support schemes and support levels.

Income-based taxes paid are represented in Figure 2.11.

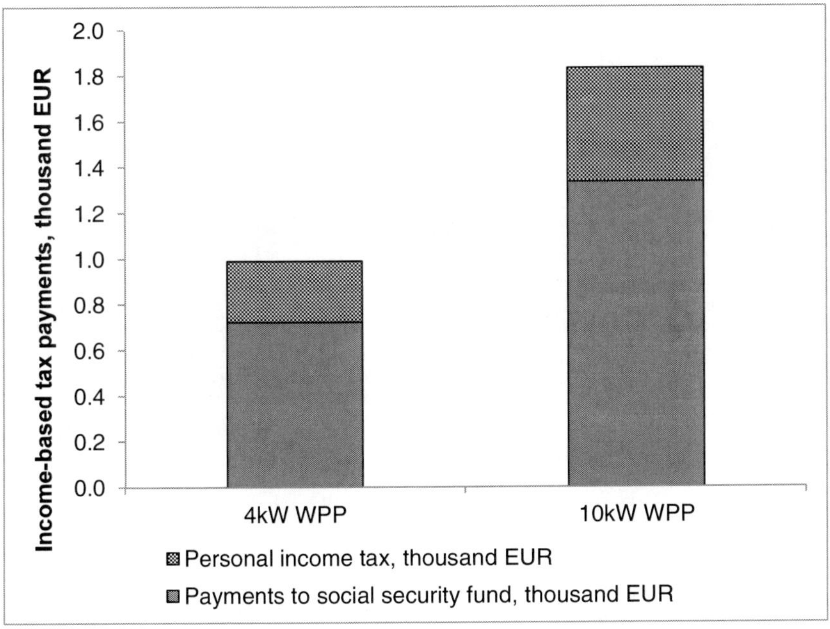

Figure 2.11. Income-based taxes of 4 kW and 10kW onshore wind turbine (author's computations).

During its lifetime, 4 kW wind turbine would generate 0.7 thousand EUR into the social security fund and pay 0.3 thousand EUR into the State/Municipal budget as a personal income tax. In case of 10 kW WPP, social security payments would reach 1.3 thousand EUR, and personal income tax – 0.5 thousand EUR.

Figure 2.12 represents distribution of requirements for domestic and foreign assets/capital.

Small-scale wind turbines are produced domestically; therefore, domestic assets/capital requirements dominate in the total asset/capital structure. Domestic assets/capital could cover up to 90% of total requirements for capital/assets. Since small-scale WPPs are domestically produced but not imported, they make a positive influence on a current account of a trade balance by the amount equal to the domestic assets/capital, which could be 8.4

thousand EUR in case of 4 kW WPP, and 15.5 thousand EUR in case of 10 kW WPP.

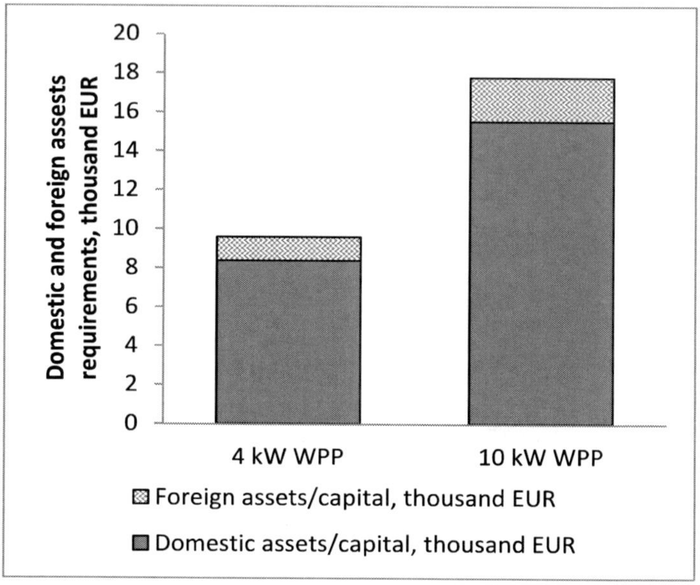

Figure 2.12. Distribution of domestic and foreign assets/capital needed for construction of 4 kW and 10 kW onshore wind turbine (author's computations).

2.5.4. Comparison of Results and Discussion Concerning Offshore Wind Power and Hybrid Power Systems Development Possibilities

Figure 2.13 provides summary of the computations. As it is seen, in many cases, investment subsidy is an attractive option for the investor. It could be used to promote wind electricity sector development in Lithuania in future.

However, quicker implementation of RES technologies is restricted by slow demand rates. The main problem is not only high initial capital costs, but also measuring of public interest, which is not included into the market transactions. Public interest is expressed not only directly through the continuous supply of energy resources in the future, the solving of social and environmental problems, but also through the implementation of EU Directives.

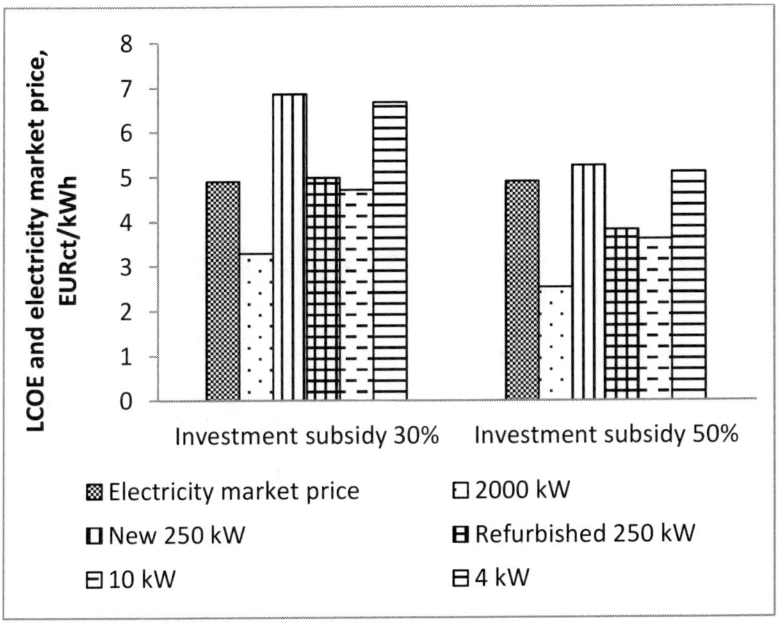

Figure 2.13. Comparison of LCOE subject to different level of investment subsidy (author's work).

There is another large source of clean energy, which is untapped yet and may become one of the most important contributors to country's energy independence in the future: offshore wind power. According to the assessments of experts, the Baltic Sea is considered as particularly suitable for offshore wind power development, and Lithuania has a good opportunity to tap this RES. First steps have already been made towards offshore wind power development in the Lithuanian Exclusive Economic Zone (EEZ), but still there are many barriers to overcome before the actual project implementation can begin.

The Law on Renewable Energy Sources (2011) states that the Government or its authorized institution must adopt necessary legal acts regulating the construction and exploitation of electric power plants in the Lithuanian territorial sea, exclusive economic zone and coastal waters. Master plan of the territory of the Republic of Lithuania must be supplemented with the part for the seawaters. Such supplement has been already introduced, but still not approved by the parliament of Lithuania.

Several reports have been made showing the potentially suitable territories for offshore wind farms in Lithuanian EEZ. They conclude that at least 1000

MW of wind turbines could be deployed in these areas. However, only an official document like an approved master plan can be a legal basis for more detailed investigations. The approval of such document is highly important, as it determines the territories potentially suitable for offshore wind power development. The second step would be the issue of permissions to explore the sea conditions at specific locations organized as a tender procedure, but, again, the description of such procedure has not been prepared yet. There are several potential investors in offshore wind farms able to implement large projects are ready to carry out the seabed and wind resource investigations at their own expense, but the main barrier at the moment is the lack of legal documents.

Experts distinguish the Baltic Sea as having high offshore energy potential and as even more suitable for wind farm deployment comparing to other waters around Europe. The advantages lie in a smaller scale of waves, less saline water and favourable seabed conditions, that is, the Baltic Sea is rather shallow and suitable for most of wind turbine foundation types.

The latest study from Kaunas University of Technology has revealed that the transmission grid in the Lithuanian coastal region must be reinforced and expanded to be able to accept the offshore wind power. Besides, the same study emphasizes that adding a significant amount of fluctuating generation capacities will force to build new power plants run on fossil fuels to balance the Lithuanian electricity system.

Moreover, the success of further offshore wind power development in Lithuania will depend not only on the solutions of technical issues but also on the stability of legal situation during the tender process, the rules of the issue of permissions in energy sector, grid connection possibilities and support scheme. Experience of other countries (UK, Denmark, Germany) has shown that feed-in tariff is the most attractive support measure bringing benefits both to state and investors.

However, despite all the challenges and barriers, it is widely agreed that Lithuania has rich untapped offshore wind resources, which in the future can become one of the main ways to secure energy supply and energy independence.

Hybrid renewable energy systems (HRES) are becoming popular for various power generation applications due to advances in renewable energy technologies and continuous rise in prices of fossil fuels. A hybrid energy system usually consists of two or more power plants run on RES. Joint use of several energy sources usually increases system efficiency and provides a better balance in energy supply.

Depending on energy demand and available local energy resources, there may be many configuration options for such system. The most popular set of a hybrid energy system is a wind turbine combined with PV modules together with joint energy storage equipment and an inverter. Investigations have shown that in Lithuania, like in most geographical regions suitable for wind power development, wind speeds are low in summer when the sun shines brightest and longest. Solar PV power plants have high seasonal energy generation variations. It has been estimated that in Lithuania, the difference between maximum energy yield in summer and minimum yield in winter may reach 10-20 times, whereas wind energy seasonal variations are smoother: maximum average wind speed in winter is 3-4 times higher than the minimum in summer. Due to the fact that seasonal variations of wind and solar power plants are reverse, and the peak operating times for these systems occur at different times of the day and year, the use of combined system results in a more reliable power supply.

The best matches of energy generation and consumption are achieved when energy system is designed according to real data of monthly domestic energy consumption. If energy yield covers maximum demand in winter, then it is practically able to fully replace electricity from the grid.

There is a number of hybrid energy systems installed in Lithuania; their installed power varies from 0.5 kW to 10 kW and more. Systems of smaller size (up to 500 W) usually find application at remote locations with specific operating equipment requiring small amounts of energy (monitoring stations, street lights, road speed control, gas service system, etc.). Larger generating capacities (up to 10 kW or more) are installed in private houses, educational institutions, companies. So far the main purpose of these installations has been to demonstrate the technical possibilities of such systems and partly to lower the electricity bills. There are several private houses with off-grid energy systems powered solely by renewable energy sources.

Currently, the net metering system is under development in the Ministry of Energy. As soon as it is approved, it should foster the implementation of small-scale renewable energy systems for both individual and commercial use. This process should consequently lead to a higher level of the decentralization of energy generation in the country, which in turn could facilitate the development of smart grid and microgrid technologies and applications.

2.6. The Significance of District Heating As Infrastructure for the Promotion of RES Demand

Renewable energy sources (RES) will have to play the main role in moving towards sustainable economic development. Talking about the perspectives of the use of solar energy, it is evident that the main market is decided by the consumer. The most important problem, determining the slow absorption process of solar energy on the part of the consumer is the lack of knowledge and organisation, deterrent amount of investments, and especially the differences between energy suppliers and users in the heating sector. The main focus of the research is what conditions are needed for heat energy that is produced by users on the basis of solar technology, and how it would be able to compete with other energy sources. District heating (DH) may be an appropriate infrastructure for implementing RES technologies both on production and demand side due to the simultaneous production of heat and electricity in combined heat and power generation plants. Conclusions have been made concerning the scale and assumptions on how solar technologies may be competitive.

Integration of RES may be more successful after reliable methodological assessment of the positive effect for solving social, economic, rural development problems in regional development context.

However, the nature of RES determines that their usage is the most appropriate for the consumer. In most cases, with rare exceptions, the utilization of RES is possible only in part with other energy resources. The worst of it is that different options are being compared as if every type of fuel would be able to provide the entire volume of supply. In this regard, biofuel is in a preeminent position, while it is relatively cheaper. Nevertheless, the substantial increase in demand of limited resources raises the price: it is the expression of the law of demand and supply, which cannot be transformed with any strategies or plans. On the other hand, biofuel is the product of the economic activity, and its renewal depends on the extent of the activity.

District heating technology is a promising tool for implementing RES technologies on the demand side and consequently achieving energy and environment policy goals. DH has various advantages compared to individual heating systems. It is usually more energy efficient due to the simultaneous production of heat and electricity in combined heat and power generation plants. However, DH is less attractive for areas with low population densities.

The importance of DH in EU would justify a more intensive use of RES applied to this type of heating system and must be considered in the design of energy policies (Cansino et al., 2011).

General characteristics of Lithuanian DH sector, its high heat market share, especially in big cities, indicate a strong position and therefore the important role in cities' energy supply. Although DH systems were developed in the planned economy, they were not always grounded economically, and networks in many cases were shortened after disconnections of collapsed industries. Initially designed pipes appeared as oversized and then resulted in high heat losses. Bearing in mind general European energy policy strategy and targets, it is evident that DH technology in Lithuania will remain as the main technology supplying energy to buildings in large cities and towns (Kveselis et al., 2011).

The encouragement of energy producers and consumers for efficient use of RES is the major goal of energy policy, as regulated by the Energy Law of the Republic of Lithuania and National Energy Strategy. Promotion of local and RES usage and energy efficiency are established in the Energy Law of the Republic of Lithuania as priority objectives of the energy sector regulation. Seeking to promote the use of RES, the Ministry of Economy has issued an order regulating energy purchase.

As IEA (2010) recently pointed out, RES will have to play a central role in moving the world towards a more secure, reliable and sustainable energy path. The greatest scope for increasing the use of RES in absolute terms lies in the power sector. Although RES are expected to become increasingly competitive as fossil-fuel prices rise and renewable technologies mature, the scale of government support is set to expand as their contribution to the global energy mix increases. National and international regulation of RES gains an important role.

The European Union is one of the most active developers of RES promotion methods and assurance measures. Its initiatives, expressed in the Directives that aim to increase the percentage of renewable energy resource in the common energy balance up to 20%, to reduce energy consumption by 20%, to reduce greenhouse gas emissions by 20% up to 2020, would lead to a drastic reduction of EU energy dependence on import (Augutis et al., 2011).

Most experience in supporting RES is offered to the electricity sector. The EU Directive 2001/77/EC required member states to increase the share of RES in electricity sector using national support instruments. In contrast, no legislative framework on the EU level was available in the heating sector before Directive 2009/28/EC was implemented. Later, following the implementation of Directive 2009/28/EC, every member state has developed its own National Action Plan that determines specific objectives for each member state in the use of RES for each sector including heating.

Major changes appeared with the introduction of new energy policy goals by adopting the RES Directive (Directive 2009/28/EC), where DH is recognized as a promising technique for reaching overall strategic energy goals: safety of energy supply by increasing independence from imported energy resources, wider use of waste energy from industries and integration of RES into energy supply infrastructures. Article 4 of the RES Directive required member states to submit National renewable energy action plans by 30 June 2010. These plans provide detailed roadmaps of how each member state expects to reach its legally binding target for 2020 for the share of renewable energy in their final energy consumption.

Necessary preparations are conducted in all EU countries by implementing this Directive. The forecasts of Lithuania reveal that biomass is the main opportunity for implementation of these plans by 2020 (Table 2.10).

It should be noted that the attention given to the potential of solar and geothermal energy is inadequately small.

Table 2.10. Projection of RES in Lithuania 2009-2020 (Ministry of Energy, 2011)

RES	2009, ktoe	2009, %	2020, ktoe	2020, %
Solar energy	0	0	1	0
Wind	14	2	99	5
Hydro energy	37	4	59	3
Biofuel	53	6	188	9
Geothermal	5	1	20	1
Biomass	763	88	1626	82
Total	872	100	1993	100

The analysis of RES use dynamics in Lithuania revealed that Lithuania has a relatively high RES potential. The main RES is biomass. However, alternative energy sources are not used sufficiently. In order to increase the usage of RES in Lithuania and to pursue the obligations to EU, it is essential to determine the causes that influence slow development of renewable energy and to identify the main RES utilization problems.

The major part of RES that is used to generate energy in DH sector consists of wood biomass. It is essential to increase the collection of logging residues and to improve the technologies of growing energy crops by substantially expanding the use of biomass resources. However, there are different opinions: the resources of biofuel are limited locally; the major part of costs consists of the collection and preparation of biofuel for use. Wood used as biofuel is usually of the kind that is suitable for the construction and production of furniture, and the price of this resource is rising because of the competition. The cost of the collection and preparation of non-merchantable wood is exceptionally high. It is illustrated by the fact that currently, a large amount of non-merchantable wood is accumulated, because it is considerably cheaper to make chips from logging than to handle, bundle and transport the residue of logging.

The use of RES for DH could be implemented by decentralizing the production of heat, solving the problems of independent heat producer's connection to DH networks, adjusting the purchase of heat energy that is made from RES and by other measures to promote the use of RES for production of thermal energy.

One of the analysed possibilities is the use of solar energy for the preparation of hot water in apartment buildings. The state supports (in accordance with the procedure, the government compensates 15% of investments) the measures for energy efficiency improvement according to the Programme for renovation/upgrading of multifamily buildings. The funding of RES (solar, wind, etc.) mechanisms is also included. In the future, there is a chance that in newly constructed houses, there will be solar collectors for the preparation of hot water, and they will use heat provided by sun. If there is a financial potential, anyone can install solar collectors. Furthermore, the other reasons are to protect the environment and to use RES and to do something new and unusual in life. Nowadays, the main and determining factor for the decision to use solar energy for preparation of hot water is limited financial resources.

Solar collectors can be installed on the roof of the building, fixed to the wall or assembled on the ground. It is important that the biggest possible amounts of solar radiation reach the surface of a solar collector. Table 2.11 shows that 50.2% of all households are centrally provided with hot water from the city (district) heating network (city 71.3%, village 2.5%) and 1.6% are centrally provided from the local boilers (city 2.17%, village 0.6%).

As indicated in the Lithuanian National Renewable Energy Action Plan, the National Strategy for the Development of Renewable Energy Sources, among other priorities of the development of energy from RES, envisages utilisation of the existing DH infrastructure and further development of the necessary infrastructure while creating conditions for the development of RES.

Table 2.11. Dwellings by type of hot water supply (% of all dwellings) (Statistics Lithuania, 2011)

	Total	Urban area	Rural area
Centralised hot water supply from the city/town (district) heating system	50.2	71.3	2.5
Centralised hot water supply from a local boiler	1.6	2.1	0.6
Individual hot water supply	31.2	20.4	55.8
No hot water supply	17.8	7.2	41.6

It is predicted that biofuel to be consumed in the DH sector should become the biggest contributor to the increase in the consumption of RES. Taking into account technological possibilities of the DH sector and the economic advantage, heat production from RES in this sector should increase by no less than 60% by 2020.

A wider use of renewable energy enables diversification of energy supply and meeting the goals of sustainable development. An overview of the present Lithuanian and EU legal regulation of the use of RES highlighted that the promotion of use of RES is among Lithuanian energy policy priorities.

It can be stated that the potential of the use of solar energy could be considerably larger in households that are in the area of the DH system.

2.7. The Use of Solar and Geothermal Energy in the System: Calculations of Efficiency Using the Example of a Particular City District Heating System

2.7.1. The Use of Solar Collectors for the Preparation of Hot Water in District Heating System

Policies and measures aiming to enhance the use of renewable energy sources (RES) are mainly driven by EU policy. The problem arises from the Lithuanian state policy, because it supports competition on the production side of district heating. Projections of RES utilization assign an unreasonably excessive role to biomass. Meanwhile, the usage of a great potential of solar and geothermal energy is significantly insufficient. The possibilities to use solar collectors and heat pumps on the supply and demand sides in district heating system were analysed in small Lithuanian towns. EnergyPRO modelling software was used for calculations; comparative analysis and Levelized Cost of Energy (LCOE) were used for detailed analysis. RES development in district heating requires large investments; however, the use of renewable resources in the district heating system would permit diversifying the fuel mix. Subsidies for investments of solar and geothermal energy technologies would allow using them in district heating system as economically attractive alternatives. A scientific problem is to define the economic background for RES support. It is essential to analyse economic support measures for the development of RES technologies in order to make them competitive with technologies of traditional fuels and to make them relevant to actual purposes, needs and possible effects. All support schemes must form a whole of uniform and systematic economy.

The main problem is the lack of integration of more diversified fuel sources, such as solar heating or geothermal heating in Lithuania.

Unique situation of the district heating occurs in Lithuania, when competition in the supply side of district heating is required from the state policy.

Present situation shows that there is no installed capacity of solar collector field or geothermal heat pumps on the district heating supply side in Lithuania. Deep geothermal sources exist in Klaipeda city in Lithuania, but this type of district heating is not competitive enough in market economy. Beginning of the use of solar collectors on the roofs of multifamily buildings has been noticed in Lithuania. Moreover, a unique apartment building with both solar collectors and heat pumps that had disconnected from district heating network in Varena town was widely broadcasted in the media. This case was chosen for a deeper economic analysis using LCoE method.

The overview of financial sources and instruments in Lithuanian heating sector revealed that biomass is recognized as a RES type, which has opportunities to gain the district heating market in Lithuania that is currently based on combustion of natural gas. However, an information gap about other RES technologies to enter district heating market is noticable. Therefore, more diverse RES sources such as solar collectors and heat pumps are analysed in this paper, especially comparing economy of RES on the supply and demand side of district heating.

Modelling is done by evaluating the heating demand in Varena town, working period of flat plate solar collectors in a year and their efficiency. Depending on the solar collector area and heat loss parameters, the programme calculated the price of hot water made by collector, the payback period of the system, NPV, IRR, etc. An average 20-year outdoor temperature and solar radiation were calculated based on available EnergyPRO yearly data. The annual heating demand is about 42.3 thous. MWh. Heat storage is available in Varena district heating network in total 1070 m^3. Interest rate of 5 per cent is calculated according to Lithuanian bank. Inflation is taken into account as European Central Bank's medium-term prognosis, that is, 2 per cent. Lifetime of solar collectors is 20 years. Operating cost for solar collectors is about 1 euro per MWh (Solar district heating platform, 2014) of produced heat, which encompasses mainly electricity consumption and maintenance.

The main scenario analysed is that solar collectors are able to cover all heating demand during the summer time in Varena town. Land area for solar collectors is available for free. Calculation tool "f-EASY (SDH)" estimated of 10000 m^2 solar collectors' area. Investments were calculated according to market situation and answers from a supplier survey. System of solar collectors that could be connected to Varena district heating network costs approximately 2.23 million euro. Calculations of the use of solar collectors showed that payback time is about 14 years. The assumption of a stable district heating price was taken into account. Revenues are calculated as heat amount from solar collectors multiplying by district heating price of 57.4 EUR/MWh.

IRR calculations have showed that investment in solar collectors is very risky, because IRR is around 8 per cent. Usually, property of district heating companies is already bonded for banks. Therefore, the interest rate is as a rule higher than 8 per cent.

As regards the situation of monopoly and insufficiently oriented pricing, outdated DH may exist without attempting to modernise the technology and reduce the supply.

In Figure 2.14, LCOE of solar collectors connected to DH system with different levels of support is shown.

The main conclusion is that without government support (such as existing 30 per cent subsidies for investment in biomass boilers or soft loan with 3 per cent interest rate) investment in solar collectors is financially unacceptable in Varena town. It should be noted that calculations are done for the city, which has one of the lowest district heating prices in Lithuania and uses mainly biomass fuel. Investments could be economically attractive just to cities with district heating price higher than the average.

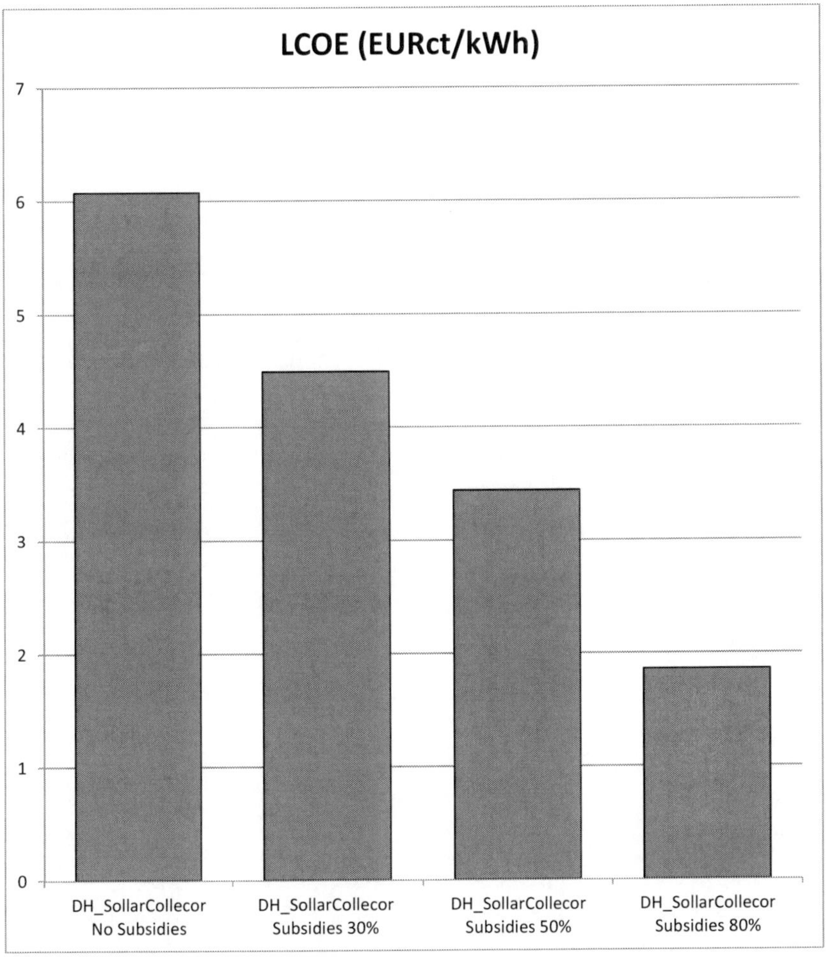

Figure 2.14. LCOE of solar collectors connected to DH system with different levels of support.

2.7.2. The Use of Solar and Geothermal Energy in the Multifamily Building and Calculations of Efficiency Using the Example of a Particular City District Heating System

The main question is competition between the price of self-produced RES energy on the demand side and the price of RES district heating on the supply side. Therefore, comparison of LCoE is shown for a solar collector and a heat pump on the demand side with different support intensity. Moreover, LCoE of renovation of apartment building is included because of the need to consume energy efficiently and compare the result of various support intensity.

A multifamily building in Varena town, where solar collectors for the preparation of hot water and heat pumps for the space heating have already been installed, is examined. This five-storey multifamily building with 40 apartments is the only one in Lithuania that already uses both types of RES technology for heating purposes.

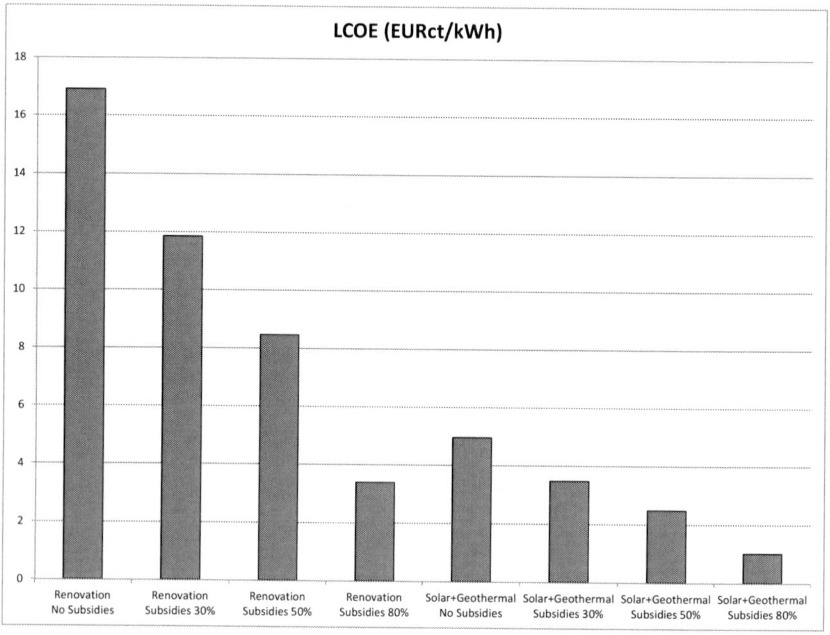

Figure 2.15. LCOE saved in multifamily house by implementing renovation or installing solar/geothermal energy sources with different levels of support.

Analysis was conducted for the renovation LCoE in multifamily building without subsidy (REN_0), with 30 per cent and 50 per cent intensity of subsidy (accordingly REN_30 and REN_50). Furthermore, LCoE of solar collectors and heat pump without subsidy (SOL_HP_0), with 30 per cent and 50 per cent intensity of subsidy (accordingly SOL_HP_30 and SOL_HP_50) are calculated. Comparison of LCoE with various intensity of subsidy reveals that in the case of complete renovation of the multifamily building, even with 50 per cent intensity savings, heating cost is higher than the existing district heating price (57.4 EUR/MWh) in Varena town. Moreover, a complete renovation requires large investments. This problem could be solved by limits of separate effective elements of renovation, such as renovation of heating system, replacement of windows, etc. On the other hand, RES on the demand side, such as solar collectors and heat pump, could compete with district heating price without any subsidy in Varena town (Figure 2.15).

If subsidy for this type of RES is given, the increase of solar collector field in Lithuania could be noticed. The similar boom of solar PV technology occurred in 2012 of high feed-in tariff for electricity generation from PV installations in Lithuania. State support has an essential role in changing the situation from moving towards cleaner energy production and diversifying fuel mix in district heating sector of both the supply and the demand sides.

Development of RES in district heating requires large investments; however, the use of renewable resources in the district heating system would allow diversifying the fuel mix. Subsidies for investments of solar and geothermal energy technologies would allow using them in district heating system as economically attractive alternatives. Moreover, if ecological, economic and social benefits were comprehensively evaluated in a long-term period, and on that basis, would be given support for consumers and suppliers, who produce energy using solar or geothermal energy, the demand for advanced RES technologies would noticeably increase.

Chapter 3

Formation of the Evaluation Principles of State Support Measures

3.1. RES Sector Financing Related Issues and Tendencies of New Investment in the RES Sector Worldwide

The growing demand for energy, commitments taken to reduce greenhouse gas emissions, long-standing energy security, energy poverty, fossil fuel depletion and other global as well as regional issues enforce humanity to search for ways how to develop their energy sectors in future. It is agreed that energy sectors should be developed sustainably, which means that, apart from other actions to be taken, it is essential to use the economically feasible potential of renewable energy sources. However, because of immaturity of (some) RES technologies, high capital cost and unsuitable pricing of fossil fuel (not in all countries externalities are considered when price for fossil fuel is set), the potential of RES is not sufficiently used. Seeking to increase the use of RES at more rapid rates, governments are implementing various RES support policies. During 2005–2011, the number of countries with some type of RES support policy more than doubled, i.e., increased from 55 (2005) to 119 (2011) countries (REN21, 2011). RES support policy and various support measures highly contributed to increasing consumption of RES. However, more and more often, scientists state that because of the above-mentioned global and regional issues, a huge new investment will be necessary in the RES sector in future. Thus, the issue of how RES sector should be financed to raise the capital and investment is of high importance. This issue is relevant both in developed and developing countries in times of slow recovery of global economy.

There are three RES sector related financing issues. They are: 1) policies that leverage increased investment in the RES sector; 2) RES sector financing channels; 3) and RES sector financing instruments. Liming (2009) suggests that instead of the analysis of policy that leverages increased investment in the RES sector, financing environment should be analysed. It is understood as favourable regulatory, legislative and policy conditions that are critical for financing RES. This issue of RES financing was widely discussed in scientific literature. However, the latter two issues are under the investigation, especially in time of economic recession and slow recovery of the economies. Thus, this chapter aims at discussing existing and possible RES sector financing channels and instruments and computing the impact of selected innovative financing instruments on energy cost.

RES financing channel is understood as the source of financing RES, i.e., it shows who is financing RES and answers the question "where do RES sector financing sources come through?" Financing instrument is determined as a delivering method of RES financing and must answer the question for what merits (i.e., usefulness) it is given the financial resources, and in what instrumental form.

Lithuanian energy represents ambitious targets and requires large investments. Lithuania besides EU objectives has national strategic energy projects, such as new nuclear power plant project, LNG terminal, electric network connection projects "LitPol Link" and "NordBalt," and all together it requires about 6.7 billion EUR. Renovation of district heating network, including modernization of CHP facilities, requires about 0.32 billion EUR (Lukosevicius, 2013). The value of building renovation programme is around 1.44-1.75 billion EUR; however, this is a long-term programme, so the annual need of investment is around 145-173 million EUR (Stankevicius, 2012). However, the gap between the objectives and available resources (i.e., financial and human resources, fossil and RES) is the main weak point, which undermines the possible fulfilment of the EU and national policy objectives. A further weak point concerns the fact that the size and the impact of integration and implementation of energy policies and energy programmes into the regional and local environment are not clear. This problem can be solved through integration of regional development resources using EU structural funds and sustainable energy objectives may be reasonably integrated in the regional development strategy in an overall vision of a sustainable development for the region (Klevas and Antinucci, 2004). However, sustainable energy policies and objectives actually are not a major issue in the development agenda of local and regional bodies. The main problem is that energy system is efficient if the investment is giving economic payback in a short term. Nevertheless, the development of RES is inconceivable without financial support so far, but the promotion of RES is possible from the attitude of total costs, including external costs, such as environmental and correction of market errors (Knapek, 2010). An appropriate assessment of economic activity can mitigate adverse effect of regional energy economy development.

Seeking to satisfy the growing energy needs and reduce greenhouse gas emissions, it is important to use more RES. However, wider consumption of RES is highly related to investment. This section of the monograph presents the tendencies of investment in the RES sector worldwide (Behrens, 2009).

New investment in the RES sector had a tendency to increase by a double digit prior to a global economic crisis, which picked up steam in 2009 (Bloomberg New Energy Finance, 2012). Highly increased volume of investment through asset finance, small distributed capacity and public markets contributed to increased volume of investment from 40 billion USD (2004) till 132 billion USD (2007). The changes in investment of the RES sector were associated with the fact that a lot of RES support policies and measures were functioning, technologies improved and their cost reduced.

The financial crisis and the deep economic recession started in 2008. However, this had only a small impact on investment in the RES sector. Volume of investment in the RES sector remained positive and reached 160 billion USD in 2008. The largest share of investment (70%) was directed to RES generation projects through the category "asset finance."

During 2010–2011, new investment in the RES sector increased. The percentage growth of investment would have been larger in 2011 if the cost of photovoltaic (further in the text PV) and wind technologies were not reduced (Frankfurt School of Finance & Management, 2012). PV module prices fell by 50% and on-shore wind turbine prices by 5–10%. In 2011, 258 billion USD were invested. The structure of investment in the RES sector was the following: 61% – asset finance, 29% – small distributed capacity, 4% – public market, 2% –venture capital, 2% – government R&D and 2% – corporate R&D. Developing economies made up 35% of this investment, and 65% of investment was made in developed economies (Bloomberg New Energy Finance, 2012).

China, USA, Germany and Italy were the largest investors in the RES sector. In 2011, they invested 51, 48, 31 and 29 billion USD, respectively. The boom in small-scale PV in Italy and Germany was influenced by the impending expiry of subsidy programmes. China and USA were leading in asset finance in 2011. 49.7 billion USD in China and 40.9 billion USD in USA were invested. Namely, India was considered as the country, which showed the highest growth rates (63%) in investment in the RES sector in 2011, followed by the United Kingdom (59%), Spain (45%) and Italy (43%). Investment in the RES sector reduced by 18% in the Middle East and African countries in 2011. The main reason for this was policy uncertainty that emerged because of Arab Spring, which influenced the postponement of some projects.

Historically, solar and wind technologies attracted the highest shares of investment. In 2011, 147 billion USD were invested in solar sector and 83.8 billion USD in wind sector. Developed countries invested more in solar sector (80% of total new investment in the solar sector), whereas developing countries invested in the wind sector (56% of total new investment in the wind sector). 10.6 billion USD are invested in biomass sector worldwide. Biofuels, small hydro, geothermal and marine sector attracted 6.8, 5.8, 2.9 and 0.2 billion USD, respectively, in 2011.

Meanwhile, in Lithuania it is believed and stated that solar technologies have almost no prospect. Although the extent of the use of wind has significantly increased during the analysed period, usually, only the major wind farms are meant.

Thus, the overview of tendencies of investment in the RES sector disclosed that investors see RES sector as of high potential. Solar and wind technologies are the most attractive for investors. The largest shares of investment go to the asset finance and small distributed capacity. Public market also plays a significant role in financing the RES sector.

3.2. The Need of Support Schemes for Energy Suppliers

3.2.1. Analysis of Market Failures

Renewable energy sources are more evenly distributed in the world compared to fossil fuels, but economical attractiveness of RES is significantly limited by a number of factors such as competition for land use, environmental restrictions and terrain constrains, intermittence of wind and solar availability. In order to improve the penetration of renewable energy sources, state support is necessary to eliminate two major energy market failures: external costs of power supply arising from pollution and benefits of RES as public goods facilitating introduction of new technologies and positive impacts on the environment. Therefore, in order to overcome these market failures and other market barriers that hinder the development of renewable energy resources, it is necessary to incorporate new market mechanisms to reduce entry cost of renewable energy, increase competitiveness in the market to encourage RES development. To achieve these goals, subsidies to renewable energy resources or obligatory quotas to the TSO/DSO or end users could be applied.

Another important market failure affects implementation of efficiency improvement measures. These measures as well as renewable energy sources are penalized by market failures associated with externalities including benefits of measures to increase energy efficiency as a public good, requiring public support in order to ensure optimal level of development in terms of overall benefits to the entire country. Even the removal of subsidies for traditional energy sources still leaves additional market barriers such as the lack of technological knowledge, differing interests of investors and consumers, and high transaction costs for individual investors. All factors mentioned above keep the cost-effective energy efficiency potentially underutilized (Brown, 2001).

Also, the market does not ensure long-term funding for research and development. New energy technologies constantly face obstacles across the whole energy innovation chain (from research and development to demonstration projects and dissemination of information and technology). The main market failures that hinder the implementation of sustainable energy development objectives are:

- Energy markets are not able to ensure the security of energy supply;
- Market underestimates the environmental and social aspects;
- Market cannot guarantee support for long-term research and development, implementation of energy efficiency measures, expansion of renewable energy sources and new technology development.

Factors Hindering Funding of Renewable Energy

Renewable Energy inexhaustibility assessment inevitably goes beyond the scope of economic theory alone: analysing what is the essence of the technical progress of delectable resources and how to transform the economic theory and policies to utilize various market imitating instruments. Studies surveyed have shown that the benefits of the assessment of renewable energy resources, based on the assumption of reduced discount rate, are suitable for the evaluation of energy utilization projects (Klevas et al., 2009).

There are no simple rules to determine the socially optimal discount rate. In fact, the choice of the discount rate in the scope of the analysis discussed remains one of the most difficulty quantified pace of data in the economic theory problems, because it brings normative questions, which go beyond the limits of the scientific discipline.

Economic theory has shown that the distribution of benefits within the selected time frame can be quantified (at least in theory) the selection of appropriate discount rate.

This issue is most closely associated with all renewable energy sources in the broader sense (i.e., starting from research to infrastructure funding necessary for RES integration into the energy networks). Existing financial risks and other barriers significantly limit growth potential of the sector.

Cost of renewable energy sources in the short and medium term is not competitive compared to conventional energy; moreover, RES projects are associated with high capital costs and low profitability level. This is the reason why most investors are not willing to invest in "high-risk low-yield" activities, or if they invest, it is done in adverse conditions. This situation is observed in developing countries, where access to finances is complicated, and the amount of financial resources is defined by the volume of targeted subsidies.

European Commission (1995) states that in the developing economies and economies in the transition, investment in the RES is complicated by underdeveloped financial markets, lack of financial institutions and available financing methods. These countries are also characterized by high political, credit, foreign exchange rate fluctuations and economic risks. In this context, an economic risk is defined by the ability to sign a favourable power purchase agreement. Liberalization of energy markets initiated the growth of independent producers, which supply energy to the grid under long-term contract. In the long-term, cash flow of power sales contract is determined by the contract signed (amount and terms of energy sale); therefore, financing of the project is possible with lower amount of private capital. Most of the independent power producers (including RES) tend to sign long-term power sales contracts. Changes in the structure of the market could be observed: a shift from long-term power purchase agreements to the participation in the Spot Market (European Commission, 2005). These tendencies complicate RES sector development by reducing possibilities to receive funding: energy sold in the short-term market cannot guarantee stable and secure cash flows therefore the financing of the project. Since investors have no guaranties that the energy from new RES facility will be bought in a spot market, where prices are determined by all involved payers (both buyers and sellers), they require that the new RES project be funded with higher share of equity, i.e., investors usually require the debt-equity ratio of less than 50 per cent. Even RES projects that can ensure long-term energy supply contracts may not be acceptable for the investor due to the disorder in the energy market. This disorder is related to the increase of Counterpart Risk in the energy market, as the energy supplier may not always agree to pay for the purchase of energy. Disorder in the deregulated energy market led to the withdrawal of some private investors (investors do not participate in the market due to the long payback period, because most investors have a short investment horizon). It leaves the public companies the main investors in the power sector. These companies in the regulated energy markets are considered to be less risky; therefore, they can even be given a higher credit rating (up to AAA). This treatment allows obtaining financing for the development of RES at the lower cost.

Legal risks – risks associated with possible difficulties in obtaining building permits, signing a fuel supply contract, getting other types of permits and approvals from environmental authorities, building inspection, etc., risks arising from the improper regulation of the energy sector (Ferderer, 1996). In the vertically integrated market, a monopolistic company may limit the use of the transmission network or may demand a higher price for the networks access. RES access to the transmission or distribution network is necessary, because energy production may be in the different location than consumption. Energy suppliers also should have access to the transmission and distribution networks.

Installation risk – the possibility that the plant will require significant increase of investment, or it will take longer to build than expected, or prove that it is not possible to install the full planned capacity. Lower performance risks are associated with the possibility that planned equipment operation efficiency will not be achieved in the first year. It is argued that these risks will go down later, when a number of similar projects have been implemented increasing available know-how. In addition to the factors mentioned above, technical factors also may influence the development of RES projects. New technology development may be influenced by the lack of local capacities and the underdeveloped network infrastructure. These reasons caused public agencies become the main investors, leaving private investment volume low.

The political risk is associated with the possibility that in the future, currently existing RES support schemes will be abandoned. For example, in Sweden, it was planned that the current RES support system of green certificate would continue only until 2010. Compared to the former system of subsidies, certification system operation horizon is longer, but still too short for proficient investment in RES technologies. Existing uncertainty affects various market participants as well as technology differently. Improvement of existing facilities requires a relatively small investment; therefore, the financing volume changes only slightly; in this way, additional funding is associated with a lower risk and cost. Investment in the construction of new generating capacity is strongly affected: due to higher financing requirements, banks tend to limit lending for the development of RES. Current slowdown in new plant design and construction of the new RES units in future can only be accelerated in the case the improvement of existing equipment options has been exhausted, and countries increase their commitments in the RES utilization. To meet these higher RES utilization commitments, further investment will have higher level of risk and therefore greater costs.

A widely accepted opinion argues that renewable energy sources are more expensive than conventional energy, but current trends in renewable energy market development show a noteworthy reduction in these costs. RES technologies are constantly improving and eventually become cheaper. Now, the cost of wind turbines and solar PV is one-tenth of the cost observed in the eighties of the 20th century. Moreover, an additional average annual cost decrease of five per cent is expected. It is no wonder that these two RES segments enjoy increased funding. For example, recent wind energy projects have been financed by issuing bonds.

Financial sector showing ever-increasing interest in RES, i.e., investment banks (such as Goldman Sachs) launched a research dedicated to the analysis of industry trends. These steps reduced insufficient dissemination of risk information about the RES sector and have facilitated RES participation for the capital markets.

Ferderer (1996) extends the list of possible risks. Risk derives from the market imperfections, inadequate regulation and risks associated with the project cost. Each risk group is analysed making a comparison between RES and conventional energy in the market. The first risk group covers RES project cost and price risks, which is divided into the following subgroups:

1. Grants to the sectors competing with RES. The World Bank and the International Energy Agency's annual budget of subsidies for fossil

fuel sector is 100-200 billion dollars. Such large volumes of subsidies can significantly reduce energy costs. If support to the RES sector is not at the same (high) level, it could lose the competitive edge.

2. Risk assessment of traditional fuel price change. Fossil fuel price change is hard to predict, but experience has shown that fluctuations in the fossil fuels price have not been important factors in planning of power sector: for a long time, fossil fuel price has been stable or slightly increasing. However, in the deregulated energy market, there is a risk of fuel price change becoming an important factor. RES technologies do not have fuel costs (excluding biomass); thus, the risk of fuel price change is avoided. While it is an advantage for RES technologies, this decreased risk has not yet been taken into account neither in economic analysis nor in technological comparison. In some regulated markets, companies are allowed to include fuel prices into the price of energy directly. For this reason, fuel price volatility risk is shifted from the producers to the consumers. In this case, investment decisions could not consider fuel price change as an important factor for project comparison.
3. Transaction costs. RES projects are generally smaller than the projects using conventional energy. In case the international financial institutions or the private sector can offer venture capital, they are usually tailored for large-scale projects and are not suitable for small businesses or small-scale projects with a total value of less than 5-10 million EUR. Furthermore, the implementation of RES projects, due to technological innovation, faces the need of additional information. This results in higher transaction costs for RES projects, which also include the use of resource assessment, site selection, permit acquisition, planning, power purchase contract and similar costs.

Market imperfections include:

1. Lack of access to financial resources. Final users or project developers often do not have access to credit for purchasing or investing in RES due to low credit ratings or the fact that the capital market is underdeveloped. For the rural households that want to install RES for their own use, micro-credit opportunities are also limited, as the proposed term of the loan is considerably shorter than the life cycle of the entire project. In some countries, the banking sector does not finance RES projects, because they are not sure whether the energy

supplier has a long-term power purchase agreement in place. Outside the OECD countries, private equity investors are looking for additional guaranties. Even taking into account sound return on investment (6-14 per cent), it is insufficient to mitigate the various risks witch may arise in many developing countries. When competing for scarce private equity funds, these facts became a distinct disadvantage and the main reason why RES in general lack funding.

2. Lack of information, commercial and technical skills. In the successful functioning of markets, its participants have the adequate skills and qualifications. The relevant information is available at low cost. However, in specific markets (such as RES), there is a shortage of experienced and qualified staff able to install, operate and maintain the RES technologies. Project developers usually do not have adequate financial, technical and business development skills. All involved sides: users, project managers and lenders have limited information describing characteristics of particular RES technology, the economic costs and possible benefits. Lack of skills and experience is associated with the increased interest rate.

The introduced classification of investments in RES has two possible streams of funding: direct (consumers savings, venture capital, groups of investors, loan on deposit or through project financing benefits) and indirect (regulatory support measures).

The key means for financing innovative technologies usually rely on the state support (capital grants, soft loans). Recently, venture capital began to actively participate in financing innovative technologies:

- Start-up capital for new companies in the research and development phase;
- Young successful business needs additional capital for further growth;
- Established business with successful market ready products requires an additional capital injection for further development.

The main methodological issue is how to deal with the confrontation between the short-term need to increase the discount rate, since it is necessary to offset a variety of risks, and long-term RES advantage, which could be achieved by reducing the discount rate. This method can provide sufficiently accurate assessment of the acceptability of each of the comparable investment decisions subject to the state's economic policies, such as tax incentives or subsidies for renewable energy.

3.2.2. Financing Instruments Helping to Compensate the Market Failures

Financing instruments for development of RES sector can be divided into two broad groups. They are: 1) governmental instruments, which are implemented through various governmental programmes; 2) and instruments provided by other financing channels (Vassilakos et al., 2003). Kalamova et al. (2011) distinguished four types of financing instruments. They are: 1) capital subsidy; 2) loan; 3) equity; 4) and mezzanine. Analysis has been performed considering classifications of financing instruments proposed by Vassilakos et al. (2003) and Kalamova et al. (2011). The initial analysis of RES sector financing instruments reveals that they do not significantly differ between countries with a different level of economy development; therefore, it is decided to discuss them in one section of the monograph.

Tax incentives are the most widely used governmental instrument, which is applied during various stages of life cycle of sustainable energy technology innovation and project development (UNEP, 2005). It was applicable in 62 out of 195 countries in 2010 (REN21 MAP, 2011). Alongside with capital investment tax incentives are a supplementary instrument to facilitate investment in the RES sector (Dusonchet & Telaretti, 2010a). Tax incentives are used, because they promote the introduction of renewable energy in many ways: 1) for renewable energy producers, they effectively increase a company's after-tax earnings; 2) for energy consumers, they reduce the cost of energy consumption, leading them to choose RES over the fossil fuels; 3) they also help to stimulate the development of a local manufacturing capacity (Jalilvand, 2012).

Jalilvand (2012) segregated 10 types of tax incentives that could be implemented to promote the introduction of RES and facilitate investment in the RES sector. They are: 1) investment tax incentives; 2) production tax incentives; 3) property tax reductions; 4) value-added tax reductions; 5) excise (sales) tax reductions; 6) import duty reductions; 7) accelerated depreciation; 8) RD & D and equipment manufacturing tax incentives; 9) tax holidays; 10) and taxes on conventional fuels.

Investment and production tax credits, import duty reductions, value-added tax reduction or exemption are a common instrument in many countries.

They are also used at the state level in the USA, Canada and Australia. Besides, in Australia, the R & D tax deduction is the major instrument fostering innovation. The aim of this tax incentive is to encourage R & D activities in Australia and to make companies internationally more competitive by encouraging the development of innovative products (KPMG, 2011). Additionally, tax allowances against taxable income of individuals and companies exist. For example, 55% tax allowance is applied for the installation of solar panels, for the production of hot water and for the replacement of winter heating systems with systems equipped with high-efficiency heat pumps, i.e., low-temperature geothermal systems in Italy. Tax incentives are also popular in developing countries. For example, in Philippines, RES projects are for seven years exempted from income tax; as well zero-value-added tax rate for RES projects is applied. Reduced rate of value-added tax for energy from RES is also effective in China (Liming, 2009). 7% of income tax is deducted for the first five years on the investment of RES equipment or technology; the purchased RES equipment qualifies for the accelerated depreciation, and income tax is deducted for 10%~20% of renewable energy stock owned by individual or company in Taiwan (Keyuraphan et al., 2012). It is worth noting that as a reaction to economic recession, a lot of countries reconsidered tax policy related to the RES sector. For example, USA extended/increased tax credits to RES sector: 1) for residential sector, investment tax credit of maximum 2,000 USD was removed, and 30% federal tax credit was approved for installation of solar and geothermal plant; 2) for business sector, the previous 4,000 USD cap for purchase and installation of small-scale wind was removed, but 30% federal tax credit was approved; 3) for facilities generating power from wind, closed-loop biomass, and geothermal resources, renewable energy production tax credit was increased to 2.1 cent/kWh for the first ten years of a renewable energy facility operation (Environmental Protection Agency of USA, 2009). Tax credits were recognized as more efficient than cash grants in the USA (Michael & Novogradac, 2011).

Capital subsidies, *grants and rebates* are typical instruments of the governmental organizations used to promote RES sector and investment in it. They function well if they are applied to market-ready technologies that are prepared to move a technology from the prototype stage to a mass production. They aim at reducing technology costs, demonstrating technological feasibility, revealing potential market barriers and increasing market penetration (Lantz & Doris, 2009). With reference to Kalamova et al. (2011), these instruments can be convertible to loans or equity when RES technology achieves commercial success. With reference to data of REN21 MAP (2011), capital subsidies, grants and rebates were applied in 42 out of 195 countries in 2010. Currently, they are used both in developing and developed countries. Nonetheless, these instruments are more popular in high- and middle-income countries, with a special emphasis on solar PV technologies. Capital subsidies for solar PV are common on the national, state, local and utility levels. They cover 30–50% of investment costs. With reference to Dusonchet & Telaretti (2010a, b), 12 western and 7 eastern EU countries use capital subsides to promote solar PV installations.

Capital subsidies *for RES sector are used for the following reasons:* 1) to reduce costs for consumers and producers by giving direct support; 2) to alter the distribution of wealth or income in desirable ways; 3) to compensate for capital market imperfections that foreclose access or discriminate against identifiable groups of business enterprises; 4) to promote competition and capital subsidies designed to offset barriers to entry; 5) the desirability of capital subsidies is attributed to perceived externalities in private sector investment. For example, capital subsidies are seen as a direct incentive for businesses to expand politically favoured activities, such as pollution control or energy conservation; 6) to promote the mobilization of private finance for renewable energy in developing countries (Fournier & Rasmussen, 1986; de Moor, 2001).

The analysis of capital subsidies, grants and rebates for RES sector in European countries showed that grants are provided by both governments and EU; as well a mix of two financing channels can be used. Grants to RES sector are provided through EU Structural Funds. Besides responding to economic recession and implementing European energy policy objectives, in 2009, European Parliament and the Council established a programme to aid economic recovery by providing Community financial assistance to projects in the field of energy. Nine offshore wind projects were granted using financial resources from this programme. Total cost of the offshore wind projects was 565 million EUR; grants from EU could not exceed 50% of the eligible costs (European Parliament, 2009). Grants are conducive for some of the projects to obtain the necessary loans from the banking sector during time of deep economic recession (European Commission, 2012).

Loans are a very important RES sector financing instrument. The majority of loans are provided by banks; however, the governmental organizations can also act as the lenders.

Olmos et al. (2012) argue that the later stages of technology development and demonstration are the predominant area, where soft loans can be successfully applied. Generally, at this stage, conventional loans from the banking sector are rarely available because of high technology development risk, the lack of immediate revenue-generation potential to repay the loan and because a small developer's balance sheet generally does not provide sufficient collateral. Thus, public sector provides soft loans or convertible loans at the pre-deployment stage of technology development. RES sector related entities use these loans for several reasons: 1) payback period of soft loan is extended; 2) low or zero interest rate is offered; 3) payback grace period can be foreseen for a soft loan (Rezessy & Bertoldi, 2010). In addition, conventional loans can be provided to finance expensive RD & D activities. For example, Risk Sharing Financing Facility, which was created by European Investment Bank and European Commission, provides loans to entities. Although Risk Sharing Financing Facility is technology-neutral, it particularly supports projects in the following sectors: 1) concentrated solar thermal power plants featuring innovative components; 2) offshore wind farms with new components; 3) and corporate RDI programmes.

Senior debts (provided in a form of credit line or loan) are also available to develop RES sector related activities and projects. The peculiarity of senior debt is that it is the first level of a corporation liabilities, which means that it is paid first, ahead of all other creditors, i.e., in the case of RES related company meets difficulties, senior debt providers are sure that they will be paid first. Thus, the advantage of senior debt, as opposed to the subordinated debt, is that it assures investor that he/she is first in seniority and is often secured by collateral. Due to the low risk, it provides the least returns to the investor, but it is a cheaper financing channel to the RES related company. Senor debts provided by public sector can offer a longer term financing than that available in local financial markets (World Bank, 2013c).

The results presented by Federal Ministry for the Environment, Nature Conservation and Nuclear Safety (2009) disclosed that the financial crisis has impacted loan market. Since economic sectors faced reduced liquidity, financial institutions appeared more risk-averse when they lent to the RES sector.

Financial institutions started lending to a company that had a strong balance sheet. Although during economic recession period official interest rates reduced, rates for RES projects increased. The increase in rates was influenced by increased borrowing spreads. Spreads for on-shore wind farms in western Europe increased to about 225 points over Libor/Euribor. Depending on technology used and other criteria, spreads were different. For example, on-shore wind projects received smaller spreads than offshore wind, and solar PV – smaller than solar thermal. Banks also shortened the loan repayment period. Although some financial institutions were prepared to lend for 15 years (loans were provided for 18–20 years in 2007), other financial institutions offered loans for 10 or less years. Financial institutions also asked for high fees and required that debt to equity ratio be reduced.

Mezzanine finance instruments are becoming important in financing RES sector, but remain little used compared with traditional loan financing. The most common instruments of mezzanine finance are: 1) subordinated loan; 2) and convertible bonds. A subordinated loan is a very common type of mezzanine finance in financing RES sector. It is determined as an unsecured loan with a lower ranking in case of bankruptcy compared to senior debt. Providers of subordinated loans receive a fixed interest rate and are ranked before equity investors should the borrower be wound up. Mezzanine finance for RES sector was/is provided by FIDEME and EUROFIDEME (EUROFIDEME, 2013). FIDEME was a 45 million EUR public-private partnership that provided mezzanine finance (namely, convertible bonds) to the RES companies in France. Financing of projects was based on the proved technologies that were verified by the French Environment and Energy Management Agency and were addressed to improving the environment. Investment made during the first generation fund FIDEME are managed by EUROFIDEME. EUROFIDEME is a Greenfield infrastructure fund, the objective of which is to receive stable returns by investing in sustainable development projects. EUROFIDEME invests in European RES electricity generated projects – in solar PV, wind, hydro and biomass sectors. The fund also invests in the experienced development companies to allow them expanding their RES asset portfolios. EUROFIDEME investment instruments are: 1) convertible bonds; 2) equity in project companies and developers with an investment amount of 3–20 million EUR per transaction. Central American Renewable Energy and Cleaner Production Facility also provides mezzanine financing in the form of subordinated debt, convertible bond and preferred shares for renewable energy for seven countries (Belize, Costa Rica, El Salvador, Guatemala, Honduras, Nicaragua and Panama). It finances up to 25% of project capital costs of RES projects (World Bank).

Bonds. As a result of financial crisis, new options raising the debt from the capital market are required. Green bonds and notes come on the scene. The application of this type of instrument is not purely new. The use of Green bonds to finance RES sector was known before the time of financial crisis. As fixed income (sometimes variable returns are set) securities issued by the governments, multinational banks or companies, they were used to raise the capital for a project that contributed to a low carbon and climate resilient economy (Croce et al., 2011).

Justice et al. (2009) and Scholer (2012) acknowledge that the attractiveness of these instruments increases in the case they are rated. Till 2011, Green bonds were issued predominantly as AAA-rated securities by the World Bank and other development banks and some other entities in order to raise capital specifically for climate change and green growth related projects. Investment-grade ratings for company's notes from three rating agencies were reduced in 2012, i.e., from Fitch till BBB-, S&P till BBB- and Moody's till Baa3.

Croce et al. (2011) indicate that there are many types of Green bonds that were issued and proposed, but Climate bond as a class of Green bonds is worth paying attention to. Climate bond is a fixed-income instrument, which allowed governments to raise capital or support the private sector in raising capital, to build RES generation and its enabling infrastructure and other. The demand for Green bonds comes from different institutional investors, such as the Danish pension fund ATP, the UN Join Staff Pension Fund and the Norwegian Global Fund; they are directed towards the retail sector; besides sovereign wealth funds, hedge funds and private equity also provided interest in the Green bonds. The World Bank issued over 2.3 billion USD equivalent of Green bonds through 39 transactions in 15 currencies. These were mostly 3–7 year, AAA rated, fixed- and floating-rate notes designed to raise capital for projects that aim to combat climate change in developing countries.

By summarizing what was said, it can be mentioned that the availability of RES sector financing instruments differs depending on the type of RES and the development stage. The variety and spread of instruments worldwide disclose that they are attractive to financial RES sector. The widespread application of instruments worldwide assures that there are actual possibilities to use these financial instruments in various countries, including Lithuania. However, currently the level of knowledge about the impact of these innovative instruments on electricity and heat energy cost, which is an important constituent part of energy price, is low. Thus the following sections deal with methodology used to assess the impact of financial instrument on cost of RES energy.

This method will be applied to assess how various financial instruments could impact RES energy cost.

3.3. Coordination of Various Promotion Forms While Avoiding Market Failures

3.3.1. The Diversity of RES Financing Forms and Analysis of Their Influence

The analysis of possible RES financing channels and instruments for Lithuania allows distinguishing loan as an instrument for financing solar technologies and further development of solar sector in Lithuania. This section of the monograph is dedicated to disclosing of how instruments provided under JESSICA and JEREMIE initiatives, as well investment subsidies are favourable to finance solar sector in Lithuania, and how application of these instruments could affect solar energy costs. The main assumptions for calculation of energy cost are provided in Table 3.1.

Table 3.1. Assumptions for assessment of RES energy cost

Parameters	Criteria	Solar PV technology	Solar collector in a small multifamily house
Technical	Installed capacity, kW	25.0	7.5
	Regression coefficient, %	1.0	0.0
	Electricity production MWh/year	24.0	–
	Heat production MWh/year	–	13.14
	Lifetime, years	23	23
Cost	Fixed cost, EUR/kW/year	5.0	1.5
	Variable cost, EUR/MWh	1.0	6.4
	Investment cost EUR/kW	2000	400

Computation of LCOE of solar energy was performed considering a discount rate of 3%, 8%, 10% and 12%. Solar PV technology and solar collectors in a small multifamily house were analysed regarding the following financing structure:

- hard loan financing – in this scenario, it was assumed that total investment cost was covered using bank loan, interest rate of which is 7%, and repayment period is 10 years;
- soft loan financing – in this scenario, it was assumed that JESSICA funds will be used. Total investment cost will be financed using soft loan, annual interest rate of which is 3%, and loan repayment period is 20 years;
- venture capital financing – in this scenario, it was assumed that JEREMIE funds will be used. Risk capital is invested for 6 years, and desired profitability of risk capital is 30% a year;
- investment subsidy financing – in this scenario, it was assumed that funds through EIF are available to finance solar technologies. Investment subsidy is 30% of total eligible cost.

The results of computation of LCOE for solar PV technology are provided in Table 3.2.

Table 3.2. LCOE for solar PV technology (discount rate 8 per cent)

	Financing method			
	Hard loan	Soft loan	Venture capital	Investment subsidy
LCOE for solar PV* EUR ct/kWh	27.78	20.38	54.82	21.16

*value added tax is excluded.

Results of computations provided in Table 3.2 showed that soft loan and investment subsidy are more preferable than hard loan. Venture capital, having high requirements for profitability, results in higher solar PV energy cost and should be applied only when commercial loans are not available. As it is seen from Table 3.2, currently LCOE of solar PV is at least by 20% higher than a maximum feed-in tariff (which was 15.06 EUR ct/kWh in 2014 Q1) set for this technology by Lithuanian NCCEP. The results of computation showed that none of the analysed financing instruments is suitable to finance solar PV technologies in Lithuania; moreover, the existing maximum feed-in tariff does not cover energy cost of solar PV. As a result, development of solar PV technologies is impeded in Lithuania. With reference to data of 2012, a maximum feed-in tariff was set very high (41.71 EUR ct/kWh in 2012), when LCOE varied in a range of 18.44 EUR ct/kWh (if soft loan was used) – 30.30

EUR ct/kWh (if hard loan was applied). Full financing of investment cost with venture capital was also unfavourable for the investor in 2012, since LCOE of solar PV in subject to venture capital financing was 42.81–63.59 EUR ct/kWh, when the maximum feed-in tariff was 41.71 EUR ct/kWh.

The results of computation of LCOE for solar collectors are provided in Table 3.3.

Table 3.3. LCOE for solar collectors in a multifamily house (discount rate 8 per cent)

	Financing method			
	Hard loan	Soft loan	Venture capital	Investment subsidy
LCOE for solar collectors* EUR ct/kWh	4.61	2.91	6.30	2.94

*value added tax is excluded.

As it is seen from Table 3.3, soft loan and investment subsidy are the most attractive financing instruments to finance solar collectors in a small multifamily house. LCOE of solar collectors in a case soft loan is 2.91 EUR ct/kWh. LCOE of solar collectors increases by 60% if hard loan is used instead of soft loan, and by 2.2 times if venture capital funds are used. Thus, it could be argued that biomass technologies could compete with solar technologies if soft loan, investment subsidy and hard loan were used to finance the solar project. Venture capital is not attractive in cities where biomass could be used. However, venture capital could be used to finance solar collectors in cities where natural gas is widely combusted.

Thus, by summarizing what was said, it could be claimed that seeking to expedite solar sector development in Lithuania, it is essential to review feed-in tariff, which is currently too low and impedes the implementation of solar PV technologies. The results of computation disclose that soft loan provided by the JESSICA initiative and investment subsidy given by EIF are instruments, which could contribute to increased consumption of solar electricity in Lithuania. Favourable conditions of these instruments impact low solar electricity cost. Thus, policy makers should consider these instruments when preparing electricity sector development plans and organizing support schemes for the development of solar PV technologies in Lithuania. Solar collectors for heat energy production are valid in small multifamily houses in Lithuania. It is rational to install solar collectors even under market conditions using hard loans provided by banks. This shows that solar collectors could compete in the district heating market even without any support. However, if they are implemented at the end user point, they could create a higher (additional) district heating cost for the rest of the customers.

3.3.2. Integration of Wind Power Plants into the Power System and Balancing Issues

In many countries, including Lithuania, wind power share in electric power systems (EPS) is growing rapidly. One of the biggest technical challenges of wind energy utilization in comparison with traditional forms of electricity generation is that the output of WPPs directly depends on the volatility of the wind. For this reason, a power system should maintain an adequate power reserve to compensate fluctuation of the wind speed. Therefore, an additional burden of electricity balancing is put on the power system, especially if the wind power comprises a significant share of the country's power system. Lithuanian wind farms produce about 5% of the total domestic consumption of electricity. WPPs with total installed capacity 223 MW are connected to the transmission grid, and with about 50 MW to the distribution grid. Experience of other countries demonstrates that power balancing could be improved if the forecast of wind speed was available for several days ahead. If a dispatcher could anticipate how much wind energy would be available in a few hours or days, it would be easier to plan the operation of the whole system.

Experience of foreign countries demonstrates that in liberal markets, reliable wind power prediction has a significant impact on wind generated electricity cost and allows wind energy competing with conventional energy sources. In Lithuania, EPS balancing is performed by transmission system operator (TSO), SC Litgrid. For the last three years, to facilitate balancing, wind farm generation forecasting service has been used. For the first year, these services were purchased from a Spanish company Meteologica, and for the last two years, this service has been provided by the Swedish company Vitec (AIOLOS model).

According to the requirements of the security of energy supply, electricity generation in the grid must match the load as precisely as possible. The maintenance of this balance is a task for the TSO. Variations of grid load during the day are sufficiently well known and predictable; they are described by the load curves. These curves are used for the determination of the next day's electricity demand and operation plan for the electricity generation plants.

Introduction of wind turbines into the grid distorts this routine, especially when wind power covers a significant share of electricity generation. Therefore, balancing of the grid becomes a more complex task. Operational power reserve is typically used for the grid balancing; it compensates the power shortage, e.g., if wind farms shut down during the storm, and the power in the grid sharply drops. Spinning reserve does not only cost, but also contributes to the higher environmental impact, as it is technically realised by burning fossil fuels. For these reasons, TSO needs a reliable wind farm power prediction few days ahead to be able to plan the operation of the electricity system, reduce the amount of operational power reserve (and consequently the CO2 emissions) as well as the grid balancing costs.

Actual power generation in a wind farm usually differs from the predictions; therefore, prediction accuracy is crucial. If all power generated from wind could replace the power generated by power plants using expensive fossil fuel, maximum savings would be reached.

According to SC Litgrid, grid balancing costs depend on the amount of balancing electricity, the difference between prices of the market electricity and balancing electricity. For example, if at a certain hour of the day it was planned to generate and sell 75 MWh of electricity, and the actual generation was 140 MWh, the difference (65 MWh) is the balanced electricity. Normally, balanced electricity is sold at a higher than market price, and purchased at a below market price. Balancing costs are directly related to the amount and the price of the balanced electricity, and they can only be reduced by either improving the wind power prediction accuracy or reducing the price difference.

SC Litgrid states that in 2012, about 75% of all purchased balanced electricity prices fell in the range of 63.7-72.4 EUR/MWh, and about 55% of sold balanced electricity prices reached 6.4-14.5 EUR/MWh. The highest balancing costs were during the warm season (March-July), and the lowest from September to February.

In 2012, balancing costs increased due to higher wind power variations, which were covered using more expensive generating capacities. In 2012, wind power balancing expenditure amounted to 3.87 million EUR. Two years of experience have shown that the biggest savings can be reached through the improvement of prediction accuracy.

SC Litgrid uses several types of wind power predictions: by 10 a.m. a one-day prediction, and by 2 p.m. a two-day prediction is provided. Then by 4 p.m., the prediction for the next day is updated, and additionally, the current day's prediction is updated every hour.

Prediction accuracy is estimated by the indicator of a normalized mean absolute error (nMAE). It is expressed in percentage of total wind farm installed capacity and is calculated using Eq. (9):

$$nMAE = \frac{|prediction - fact|}{capacity} \tag{9}$$

Investigations have shown that an average two-day prediction error reached 11.04% in 2012 (see Figure 3.1). Average error estimate of the one-hour predictions is as low as 6.46%.

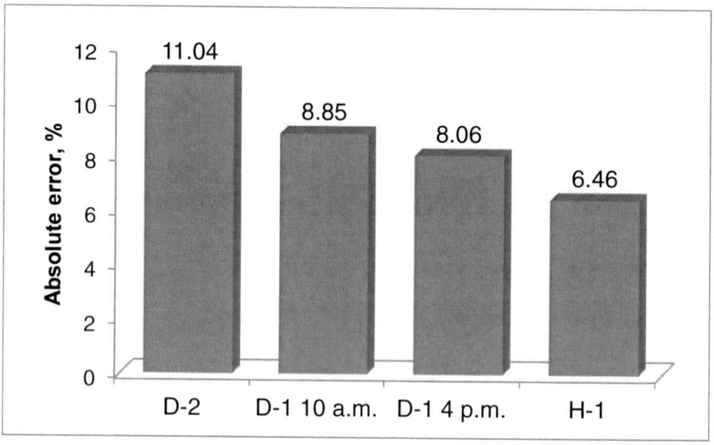

Figure 3.1. Error estimates for different types of wind power predictions in 2012 (D-2 – prediction two days ahead; D-1 10 a.m. – one day ahead prediction at 10 a.m.; D-1 4 p.m. – one day ahead prediction at 4 p.m.; H-1 – prediction one hour ahead).

Calculations have proved that balancing costs are directly related to the wind power prediction error values. Analysis shows that the decrease of wind power prediction error by 1% results in the reduction of balancing costs by around 10%. It should be noted that balancing electricity price depends on the balancing market situation as well; therefore, the development of Lithuanian balancing market will have influence on the shape of relation between the balancing price and prediction error. For example, the use of the new Kruonis HPSP unit for wind farm power balancing could reduce balancing costs. There are more possible options, but they need detailed calculations, which are beyond the purpose of this monograph.

SC Litgrid has estimated that balancing costs of wind farms with total installed power 223 MW would decrease by 43 thousand EUR if wind power prediction annual average error were reduced by 1% (Feasibility study on the integration of renewable energy power plants to the 330-110 kV transmission grid till 2030, 2014).

If Lithuania had implemented the plans provided in the Law on Renewable Energy Sources (2011) and increased the total installed wind farm power to 500 MW, yearly grid balancing costs would rise up to approx. 9.27 million EUR. This estimation should be considered when planning the funds of the Foundation of Public Interest Services (if such RES support scheme remains in the future).

Following the changing requirements in the electricity market, SC Litgrid has taken measures, and since February 2014, company is using results of a "one day-ahead" wind power predictions, which are updated two times a day. Also wind power prediction results have been supplemented by an additional source from another model run in Denmark. Results of this improvement are expected to be clear in the beginning of the year 2015.

Considering future prospects, offshore wind farms are expected to be a more stable source of wind power, this technology is promising in terms of power prediction errors, but so far, this idea is in the development stage. Hybrid small-scale renewable energy systems using wind power technologies also would add some accuracy to the predictions as compared to stand-alone wind turbines. This technology is relatively new in Lithuania and is to be overviewed in the nearest future.

3.3.3. Financing Channels Depending on the Life Cycle of the Development of RES Sector

Bloomberg New Energy Finance (2012) distinguished four stages of life cycle of RES sector development. They are: 1) technology research; 2) technology development; 3) equipment manufacturing; 4) and roll-out, which correspond to the stage of deployment of renewable energy generation project.

With reference to Bloomberg New Energy Finance (2012), RES sector is financed using both private and public financial sources and mix of financial resources. Based on the global experience, it could be argued that technology research stage is traditionally the principal focus of public financing. Aiming at reducing risk for RES technology researchers, technology research stage is mainly financed by the governmental and other public entity grants and subsidies, which are not refundable. Bloomberg New Energy Finance (2012) agrees with conclusions of Canadian Renewable Energy Alliance (2006), Justice et al. (2009) that nowadays, there is a great variety of financing channels, through which the RES sector can be financed. However, Bloomberg New Energy Finance (2012) emphasizes that financing channels highly depend on the stage of the life cycle of the development of the RES sector.

When RES technologies move out of the research to the development stage, the amount of public financing reduces. Venture capital and privative equity come into the place. Because of high level of risk and dismissal of the requirements held by the financial institutions for receiving loans, in early-stage of technology development, RES technology developer is not capable of receiving financing from credit/debt markets. Instead, they use financial resources of venture and private equity capitalists. These capitalists obtain equity shares in the start-up and fast growing RES technology developing company and start playing a significant role in the management and technical aspects of it, including obtaining a seat on the board (UNEP, 2005). Although venture capital is a subset of private equity, some differences between financing channels are noticeable. The main are provided in Table 3.4.

Table 3.4. Comparison of venture capital and private equity (Mergers Inquisitions, 2013)

Criteria	Private equity (PE) capital	Venture capital (VC)
Company Types	PE companies buy companies across all industries	VC investors are focused on technology, bio-tech, and clean-tech
Per cent acquired	PE companies almost always buy 100% of a company in leveraged buyouts	VC investors only acquire a minority stake, less than 50%
Size	PE companies make large investments, at least 100 million USD up to tens of billions for large companies	VC investors investments are much smaller, often below 10 million USD for early-stage companies
Structure	PE companies use a combination of equity and debt	VC investors use only equity
Stage	PE firms buy mature, public companies	VC investors invest mostly in early-stage, sometimes pre-revenue, companies

Venture capital comes from several types of investors: 1) institutional; 2) and high net worth individuals. The first investor is known as formal venture capitalist and is a legal entity. He/she mainly invests in the non-listed newly created companies and does not play in safe. The second one is informal venture capitalist and is known as natural entity or "business angel." He/she is rich and is ready to use his/her financial resources to finance risky investment. Together with the invested capital, "business angels" share their experience, knowledge and relations. Since "business angels" are rich individual investors, they fulfil the gap between individual financing possibilities and institutional venture capital. The advantage of this type of financing channel is that "business angels" take more flexible decisions compared to institutional venture capitalists and react to changes in the surrounding environment more quickly. Although venture capitalists provide early-stage financing, they can also be a channel providing expansion financing and acquisition buyout financing. In their analysis, Hofman & Huisman (2012) found that majority of venture capital and private equity in the field of RES sector focus on the seed and start-up capital (i.e., the early-stage financing) (35%), expansion financing (42%) and other financing (23%).

Private equity investment is used to trigger pre-deployment (commercially attractive RD & D) carried out by small entities for several reasons. RES sector related entities choose this type of investors due to several reasons. Unlike debt, equity does not create financial distress. This is due to the fact that payments to private equity owners depend on the success of RD & D. Besides, private equity provides small entities with collateral they need to receive a debt. However, some entities may want to avoid the influence of new equity providers, since new equity has large cost for existing equity owners, i.e., the value of their shares decreases. For example, large enough firms try to avoid new equity funds. Instead, they try to use loans, since they can receive more favourable interest rates in the lending market.

The main reasons for private equity capital and venture capital to enter the RES sector are various RES sector supporting policies implemented by governments. The results of Bürer & Wüstenhagen (2009) and Hofman & Huisman (2012) disclosed that feed-in tariff was the most popular policy among venture capitalist and private equity investors during economic growth and recession period worldwide, which had impact on their decision to invest in the RES sector. Comparing the results of 2007 and 2011, scientists observed that the importance and popularity of most policies decreased during economic recession period. Significant decrease was found in trade-based mechanisms, green quotas and certificate trading as well as CO2 emissions trading.

Economic recession period and followed cuts of subsidies in European countries influenced the popularity of several RES policies; however, feed-in tariffs were still the most important RES support policy among venture capitalists and private equity investors, since they provide stable incentives for investors.

Financing of renewable energy generation projects is also very important and is called "asset financing." Asset financing means that money is invested in RES generation projects, using loans and equity capital (Bloomberg New Energy Finance, 2012). As it was emphasized by Kalamova *et al.* (2011), financing the construction of renewable energy generating facilities involves a mix of equity investment from the project owners and loans from the banks (called private debt) or capital markets (called public debt, which is raised through issuing bonds). Mergers and acquisition transactions in the RES sector usually involve the sale of generating assets or project pipelines, or sale of companies that develop or manufacture technologies and services.

The analysis of RES sector financing channels showed that there is a great variety of RES sector financing channels worldwide. Financing channels differ between countries; they depend on the level of country's economic development, type of RES and the stage of life cycle of RES sector development.

Chapter 4

ANALYSIS OF SHORT-TERM RES SUPPORT MEASURES (UNTIL 2020)

4.1. SUPPORT MEASURES FOR THE DEVELOPMENT OF RES IN LITHUANIA AND THEIR ANALYSIS

4.1.1. Support Measures for RES, Set Out in Directive 2009/28/EC

Directive 2009/28/EC on the promotion of the use of energy from RES, amending and subsequently repealing Directives 2001/77/EC and 2003/30/EC publishes that the EU member states support the development of RES in electricity, heating and cooling, and transport sectors. The purpose is to increase the RES share by 2020 in the structure of gross final energy consumption of these countries to the level defined by the Directive. Moreover, at least 20 per cent of the share of RES should be reached on the EU level. EU-27 member states ought to use various support schemes for this purpose. It should be noted that mandatory national indicators of the RES share of Member States that are laid down by Directive were determined regardless of the availability of RES in EU Member States. In order to facilitate the implementation of the objectives on the EU level, the EU Member States are allowed to use following cooperation mechanisms (Directive 2009/28/EC):

- joint support schemes;
- statistical transfers between Member States;

- joint projects between Member States;
- joint projects between Member States and third countries.

Cooperation mechanisms are not applicable for the implementation of RES share target in transport sector. The following is a brief introduction of the separate cooperation mechanisms.

Joint support schemes (JSS). National RES targets of Member States might be reached by joining or partly coordinating their national support schemes. JSS mechanism allows a certain amount of RES energy, which is produced on the territory of one participating Member State, including the national target of another participating Member State. Inclusion is permitted if the following conditions are met:

- If the Member States participating in the JSS made a statistical RES energy transfer from one Member State to another Member State;
- If the Member States participating in the JSS set up a distribution rule and notified EC no later than three months after the end of the first year in which it takes effect.

Statistical transfers between Member States. Statistical transfers mean that RES could be used in one country and then virtually transferred to another Member State, i.e., Member States could agree on a statistical transfer of specified amount of RES.

Joint projects between Member States. Two or more EU Member States may cooperate on all types of joint projects that are related to the production of electricity, heating or cooling from RES. That cooperation may involve private operators.

Joint projects between Member States and third countries. One or more Member States may cooperate with one or more third countries on all types of joint projects that are related to the production of electricity from RES. Such cooperation may involve private operators.

In order to implement requirements of Directive 2009/28/EC, the EU Member States may provide support schemes and cooperation mechanisms. The cooperation mechanisms are new support measures for the RES objectives; on the other hand, EU-27 Member States have long-term experience of the application of support schemes. Promotion of RES experience in EU countries and success stories are summarized in the next chapter.

4.1.2. Promotion of RES Experience in EU Countries

Sectorial Support Schemes RES EU-27

Directive 2009/28/EC requires EU Member States to support the RES energy in all sectors of the economy, i.e., electricity, heating, and transport sectors. Different support schemes are used by the EU-27 Member States by carrying out this obligation for the promotion of RES development in electricity, heating, and transport sectors.

Support schemes of RES in electricity sector are promoted in EU-27 Member States by applying direct and indirect price support schemes. Green certificate, feed-in tariff, and feed-in premium are attributed to the group of direct price support schemes. Tax incentive, investment subsidy, and support for research are support measures that are attributed to the group of indirect price support schemes. Table 4.1 shows support measures that are used in EU-27 Member States for the promotion of electricity produced from RES.

Table 4.1. Support schemes for the promotion of electricity produced from RES

	Feed-in tariffs	Feed-in premiums	Quotas	Investment subsidies	Tax incentives	Fiscal measures	Auctions
Number of EU Members	20	9	6	12	10	8	5
Lithuania	V	-	-	V	-	-	V

As shown in Table 4.1, feed-in tariffs are differentiated according to the types of RES, the used technology, the installed capacity, the location of installed power plant, and other criteria. Feed-in tariff is the most widely used support measure for the development of electricity produced from RES in the EU Member States. It should be noted that fixed feed-in tariff levels between Member States are very different. Feed-in tariffs are periodically reviewed because of the evaluation of various factors (necessary investments in change of RES technologies, inflation rate, interest rate, market price of electricity, etc.). Some Member States (such as Germany) have introduced a fixed percentage reduction of feed-in tariff level over time. Feed-in tariffs are calculated in accordance with the approved methodology. For example, in Portugal, the feed-in tariff rate is determined taking into account the avoided external costs. Producers of electricity that is generated from RES receive a monthly fee for the electricity, which is calculated according to the formula containing evaluation of factors such as:

- Capacity of device, which reflects the investment in traditional electricity device if it were built in place of the RES unit. This component is a constant (fixed);
- Costs of generated electricity in hypothetical traditional device. This component is variable and shows how much costs per 1 kWh are;
- Avoided costs of CO_2 emissions because of the use of RES;
- Peak and non-peak electricity production, reflecting how the devices produce electricity during the day and night;
- The inflation rate;
- Avoided loss of electricity in networks by using RES.

In some countries (such as Denmark, Finland and the Netherlands), the main support scheme is feed-in premium, which is characterized by the fact that electricity is sold at the predominant market price, and the producer will receive additional premium above the market price for the sold electricity. Feed-in premium amount can be variable or fixed. Fixed feed-in premium is used in Denmark, while in the Netherlands, variable feed-in premium is used. Feed-in premium in Holland depends on the base price of electricity and the correction of feed-in premium, i.e., on the market price. Eq. (10) below is used in the Netherlands as a formula for feed-in premium:

$$\textit{Feed-in premium} = \textit{Base price} - \textit{Correction of feed-in premium} \tag{10}$$

Base price is fixed and guaranteed for the entire period of support. It is differentiated by technology. The market price is changing every year; therefore, it has an impact on the annual feed-in premium amount. For example, when the market price is increasing, feed-in premium is decreasing.

In Finland, feed-in premium scheme has come into operation only since 2011. Before this period, electricity produced from RES was supported through indirect price support schemes, such as investment subsidies and tax incentives.

In some countries (the Czech Republic, Spain, Slovenia and Slovakia), feed-in tariffs and feed-in premium schemes are used for support of electricity from RES. This means that electricity producers can to choose which support scheme they would like to use.

Only six EU Member States (Belgium, Italy, Poland, Romania, Sweden and the UK) use *green certificate-based quota systems* for support of electricity from RES. Electricity is sold at the predominant market price in this support scheme. In addition, electricity producers receive revenue from the sale of green certificates, which indicate that the electricity is produced from RES and might be supported.

Investment subsidies and tax incentives in many countries are support measures for the promotion of energy from RES.

For the promotion of RES in heating sector, only indirect price support measures are used, such as tax incentives and investment subsidies from the EU Structural Funds and national funds of the country.

Table 4.2 shows support schemes of RES promotion for the heating sector in the EU-27 countries.

Table 4.2. Support schemes for the promotion of RES in district heating sector

	Investment subsidies	**Tax incentives**	**Fiscal measures**	**Legal obligation**
Number of EU Members	24	9	6	3
Lithuania	V	-	V	-

As can be seen from Table 4.2, in addition to the tax incentives and investment subsidies, the use of RES is promoted by the obligation to use RES in the building sector. This is related to the Directive 2009/28/EC provision "... Member States shall introduce in their building regulations and codes appropriate measures in order to increase the share of all kinds of energy from renewable sources in the building sector... ." The latter Directive requires that "...by 31 December 2014, Member States shall, in their building regulations and codes or by other means with equivalent effect, where appropriate, require the use of minimum levels of energy from renewable sources in new buildings and in existing buildings that are subject to major renovation. Member States shall permit those minimum levels to be fulfilled, inter alia, through district heating and cooling produced using a significant proportion of renewable energy sources... ." Moreover, the Directive states that "... Member States shall ensure that new public buildings, and existing public buildings that are subject to major renovation, at national, regional and local level fulfil an exemplary role in the context of this Directive from 1 January 2012 onwards. Member States may, inter alia, allow that obligation to be fulfilled by complying with standards for zero energy housing, or by providing that the roofs of public or mixed private-public buildings are used by third parties for installations that produce energy from renewable sources...." Article 13 paragraph 6 emphasized that "...Member States shall promote the use of renewable energy heating and cooling systems and equipment that achieve a significant reduction of energy consumption. Member States shall use energy or eco-labels or other appropriate certificates or standards developed at national or Community level, where these exist...."

As can be seen from Table 4.2, the obligation to use RES in building sector is found in Spain, Portugal and Germany. In Spain, the obligation to use RES in building sector was set back in Technical Building Code approved by the Royal Decree 314/2006 in 2006 March 17. This Code includes an obligation to cover 30-70% of the domestic hot water demand by means of solar thermal energy. The obligation is set for new or renovated buildings, regardless of their intended use. The Code sets out the minimum obligations, the amount of which depends on the group of hot water demand and climatic zones. Local or regional authorities, taking into account the availability of solar thermal energy, have the right to tighten obligations. In Portugal, obligation to use RES in the building sector was established by Decree 80/2006. This Decree sets out the obligation to install in buildings at least 1 m^2 of solar thermal panels per capita (restriction is up to 50% of the roof surface), and use this energy for the preparation of domestic hot water.

4.1.3. Germany's Experience

Electricity sector. On April 1, 2000, entry into force of the Renewable Energy Act recognized that the feed-in tariff is the main support measure for the electricity produced from RES. Thus, supporting electricity, its share is expected to increase from 35% (2020) to 80% (2050). Two institutions are responsible for the feed-in tariff measure; they are: the Federal Ministry of Environment and the Federal Network Agency. Federal Ministry of Environment Commission has to regularly review and evaluate Renewable Energy Act. Following the entry of the last amendment of the Act into force, from January 1, 2012, producers of electricity generated from RES can choose between feed-in tariff and feed-in premium. Feed-in premium aims at that producers of electricity generated from RES would be more focused on trade based on market principles. Feed-in premium is calculated as the difference between feed-in tariff and the average monthly spot-market price of electricity.

Electricity producers have the option to change the type of support by replacing feed-in tariff by feed-in premium, or vice versa. This can be done every month. Until 2014, feed-in premium would be mandatory for all operators who use biogas, when the plant capacity is more than 750 kW. In the amendment of the Act, it is also stated that solar PV technology operators can get a feed-in premium for the electricity, which is consumed for their own needs. If the amount of consumption is higher than 30% of the total production, then feed-in premium is increasing. This provision was adopted in order to change consumption, i.e., to reduce peak consumption. Table 4.3 shows the feed-in tariffs and feed-in premium rates.

Table 4.3. Feed-in tariffs and premium system in Germany (Ecofys et al., 2011)

Technology	**Support 2011 (€/MWh)**	**Support 2012 (€/MWh)**	**Reduction rate of the tariff**
Hydro	34.3 (>50 MW)–126.7 (<500 kW)	34.0 (>50 MW)–127.0 (<500 kW)	1% (2013 and later)
Onshore wind power plants	90.2 (first 5-10 years, depending on the profitability; premium 4.9-9.8)	89.3 (first 5-10 years, depending on the profitability; premium 4.8–9.8)	1.5% (2013 and later)
Offshore wind power plants	150.0 (first 12 years)	150.0 (first 12 years)	7% (2018 and later)
Biomass (including biogas)	76.3–114.4 (premium 19.6–107.8; higher premium to biogas)	60.0–143.0 (premium 25.0–80.0; higher premium to biogas)	2% (2013 and later)
Sewage and landfill gas	59.8–87.3 (premium 19.4)	58.9–86.0 (premium 10.0; 20.0; 30.0)	1.5% (2013 and later)
Solar PV	feed-in tariff *(direct use)*	feed-in tariff *(direct use)*	1.5-24% (2013 and later, depending on capacity)
<30 kW	287.4 (123.6–167.4)	244.3 (805.0–124.3)	
30-100 kW	273.3 (109.5–153.3)	232.3 (685.0–112.3)	
>100 kW	258.6 (94.8–138.6)	219.8 (560.0–998.0	
>1000kW	215.6	183.3	
Technologies installed on the ground	211.1–220.7	179.4–187.6	
Geothermal	142.1–196.0 (premium > 10 MW 29.4–68.6)	250.0 (premium 50.0)	5% (2018 and later)

Feed-in tariffs are differentiated according to two criteria, such as used technology and the size of the project. Support is not provided for biogas plants with an installed electric capacity of more than 5 MW_{el} and based on the use of waste water and the use of landfill gas, as well as biomass plants larger than 20 MW_{el} capacity and ground installed solar PV in cultivated areas. Feed-in tariff system is not financed from the state budget. The total cost of the system is distributed to end-users. Feed-in tariffs are provided for a period of 20 years from the equipment installation on the market. For the modernized hydropower plant case, feed-in tariff is only valid for 15 years.

Feed-in tariff amount is differentiated depending on the use of innovative technology, sustainability criteria, consumption of electricity for their own needs (auto consumption), and efficiency of the technology.

Feed-in tariff and feed-in premium for new installations every year is reduced by a fixed percentage according to the on-going process of technological learning. Reduction rate for the tariff varies for different technologies. Reduction rate of feed-in tariff for solar PV technologies was introduced in 2009 (tariffs were reduced in 2010). Reduced feed-in tariff for the offshore wind power plants will be applied from 2015 and 2018, depending on whether the plant was built before or after January 1, 2012. All new geothermal power plants installed in 2012 and later will get a constant feed-in tariff until 2018.

Heating sector. The Act on the Promotion of RES in the Heat Sector that entered into force on January 1, 2009 provides the obligation for homeowners to use RES for the production of heating and cooling in new buildings with a larger than 50 m^2 area (Act on the Promotion of Renewable Energies in the Heat Sector, 2008). The Amending Act of the Act on the Promotion of RES in the Heat Sector entered into force on May 1, 2011 (Act on the Promotion of Renewable Energies in the Heat Sector, 2011), which meets the requirements of Directive 2009/28/EC. This Act states that RES must be used not only in new buildings, but also in renovated public buildings. RES must be used if the building area is more than 50 m^2. The Act has ordered a mandatory share of RES in heating and cooling. Depending on the type of RES (technology), a required share of RES in heating and cooling production ranges from 15% to 50% of newly constructed buildings, and from 15% to 25% of renovated public buildings (Table 4.4).

The Act emphasizes that annually, about 500 million EUR from the state budget will be given for the support of RES development in heating and cooling sector during the period 2009-2012.

Table 4.4. Mandatory share in heating and cooling that is produced from RES (Ecofys et al., 2011; Act on the Promotion of Renewable Energies in the Heat Sector, 2011)

RES type, technology	Mandatory share of RES
Solar	15% in new building and renovated public buildings
Geothermal energy, heat pumps, solid and liquid biomass	50% in all newly built buildings, and in 15% renovated public buildings
Biogas	30% in all newly built buildings, and in 25% renovated public buildings

Responsibility for the implementation of support is given to the Ministry for the Environment, Nature Conservation, Building and Nuclear Safety and the Ministry of Finance. These Ministries are implementing Market Incentive Programme for Renewable Energies on the Heat Market. Investment subsidies and soft loans with long-term low fixed interest rate can be obtained under this programme. In 2011, support of 380 million EUR from the state budget was allocated for the programme.

Market Initiatives Programme for the use of RES in the heating market is divided into two parts:

- Investment subsidies and grants;
- KfW's Renewable Energy Programme.

Table 4.5. Investment subsidies **by categories of supported technologies (BAFA, 2012)**

Technology	Support
Solar collectors:	
– up to 40 m^2	1500–3600 EUR
– 20–100 m^2 collectors that are used in multifamily buildings and large non-residential buildings	3600–18000 EUR
Biomass	
– pellet boilers	1400–3600 EUR
– wood chips boilers	1400 EUR
– firewood boilers	1400 EUR
Heap pumps	1300–12300 EUR

Subsidies for solar thermal energy, biomass technologies and heat pumps, which can be installed in 1 or 2 family houses, multifamily buildings, commercial and public buildings, significantly increased according to the data from August 2012 (Table 4.5).

Certified products must be installed if there is a need for support. For example, solar collectors must be certified in accordance with EN 12975.

Large equipment that uses RES in the heating sector can benefit from using the KfW support. It is given as a soft loan with long-term low fixed interest rate. Financing can cover up to 100% of the total eligible costs. The maximum loan is 10 million EUR. Small entities receive more favourable interest rates. KfW supports solar thermal technologies (collector area of 40 m^2) installed in multifamily buildings and commercial buildings, biomass boilers, high-efficiency combined heat and power plants (<2 MW), and others.

4.2. Lithuanian Experience in Financing RES Sector

4.2.1. Existing RES Financing Channels and Instruments

Although the review of experience of developing and developed countries showed that RES sector worldwide could be financed using various financing channels and instruments, experience in Lithuania discloses that thus far, traditional financing channels and instruments are used to finance the RES sector in Lithuania. The main reason for this is that financial institutions have little (some of them even no) experience in financing RES sector; as well, there is no stability in RES sector regulations, which increases risks for the financial institutions. Nonetheless, financial institutions are actively seeking for new niches. Thus, this section of the monograph analyses the RES sector financing channels and instruments in Lithuania.

Electricity Sector

Feed-in tariff *system*. RES support policy in Lithuania was established in 2002, when the application of feed-in tariff system for electricity produced from RES had just begun. It set a fixed price for produced RES electricity. The feed-in tariff was higher than electricity market price. Thus, RES electricity generators started receiving subsidy for the service provided. *This subsidy was paid by the end electricity consumers.* Feed-in tariffs were revised, however, non-periodically. Feed-in tariff system was revised in 2007, 2008, 2009, 2011 and 2013. Considering the inflation rate and other criteria, feed-in tariffs increased. Approved tariffs promoted RES electricity production. Since 2002, RES electricity generation volume increased more than 20 times from 41 GWh (2002) to 831.76 GWh (2012). Electricity production volume in wind power plants increased the most, and with reference to data from 2012, wind electricity covered 64.6% of the structure of RES electricity. Since 2010, feed-in tariff was approved for electricity from solar energy. Guaranteed feed-in tariff was applied for 2.4 GWh of electricity in 2012.

According to the Law on Renewable Energy Sources, which was implemented in 2011, feed-in tariff was provided in accordance with the installed capacity. During the first year of implementation of the Law, fixed feed-in tariff was paid for RES electricity, which was produced in the power plants of capacity less than 30 kW. Feed-in tariff was differenced in accordance with the type of technology. Legally, there was no regulation for the size of capacity, for which feed-in tariff could be paid. The difference between feed-in tariff and electricity market price was paid to RES electricity, which was produced in the power plant of more than 30 kW. Legally, it was determined that subsidy would be paid for generators that receive promotion quota. The total promotion quota for wind was 260 MW, biofuel – 230 MW, biogas – 75 MW, hydro – 14 MW, and solar – 10 MW. Promotion quota was allocated in the auction, which was organized by the National Control Commission for Energy and Prices (*further in the text* NCCEP). NCCEP also announced the maximum feed-in tariff for each technology, but the producers provided proposals for a price that did not exceed the maximum feed-in tariff. Subsidies to solar technologies seemed to be too high and regulations too low; therefore, such conditions generated a boom in solar sector. Provision of support was suspended. In 2013, the Law on Renewable Energy Sources was updated. According to the latest amendments, fixed feed-in tariff is paid for RES electricity, which is produced at the power plants of capacity less than 10 kW and more than 10 kW. Currently, promotional quotas are set for investors of both groups (up to 10 kW and more than 10 kW). Promotional quotas and feed-in tariffs are set during the auction, which is organized by NCCEP. Feed-in tariff is provided for 12 years since the license to produce electricity has been issued. Table 4.6 provides information on the average size of subsidy for RES electricity.

As it is seen from Table 4.6, maximum feed-in tariffs for wind electricity, when power plant is connected to the transmission system, are by 7.1% higher than the actual feed-in tariff agreed with the auction winner. This agreed feed-in tariff was by 2 EUR ct/kWh higher than the average electricity market price in 2013.

Table 4.6. Feed-in tariff for electricity from RES

<table>
<tr><th>Technology</th><th>Maximum feed-in tariff set by NCCEP in 2011, EUR ct/kWh</th><th>Maximum feed-in tariff set by NCCEP in 2012, EUR ct/kWh</th><th>Maximum feed-in tariff set by NCCEP in 2013, EUR ct/kWh</th><th>Actual feed-in tariff received by winners of the auctions, EUR ct/kWh</th><th>Average electricity market price in 2013, EUR ct/kWh</th></tr>
<tr><td>Wind, connected to transmission network</td><td>8.7</td><td>8.1</td><td>7.5</td><td>7.0-7.1</td><td>5.1</td></tr>
<tr><td>Wind, connected to distribution network</td><td>8.7</td><td>10.4</td><td>9.3</td><td>7.2-10.4</td><td>5.1</td></tr>
<tr><td>Hydro</td><td>7.5</td><td>7.8</td><td>7.0</td><td>7.8</td><td>5.1</td></tr>
<tr><td>Biogas, processing of biodegraded waste</td><td>8.7</td><td>16.8</td><td>13.9-14.8</td><td rowspan="2">13.6</td><td>5.1</td></tr>
<tr><td>Biogas, using landfill gas</td><td>8.7</td><td>16.8</td><td>9.6-9.8</td><td>5.1</td></tr>
<tr><td>Biomass</td><td>8.7</td><td>13.0</td><td>9.8-13.0</td><td>9.7</td><td>5.1</td></tr>
<tr><td>Solar</td><td colspan="3">Depending on the capacity, feed-in tariff varies in a range of 13.9–52.10 EUR ct/kWh</td><td>N/A</td><td>5.1</td></tr>
</table>

In case of wind connected to the distribution network, agreed feed-in tariff was higher than average electricity market price by 40–104%. Biogas example discloses that depending on the technology, agreed feed-in tariff is 2–3 times higher than electricity market price.

Environmental Investment Fund. RES electricity generators, who do not participate in the feed-in tariff system, can apply for the financing provided by Environmental Investment Fund (*further in the text EIF*). Investment projects under the EIF programme are financed using financial resources from the state budget, i.e., by using 30% of the taxes levied on the environment pollution. These financial resources are committed and provided to beneficiaries in a form of subsidy. The maximum subsidy to one beneficiary does not exceed 199.84 thousand EUR for three years. Subsidy covers up to 80% of total eligible cost (Environmental Investment Fund, 2013).

In addition to EIF programme, EIF implements Climate Change Special Programme (*further in the text* CCSP). CCSP is financed through financial resources received from selling emission allowances and from voluntary financial resources provided by legal and natural persons. Promotion of RES utilization and implementation of environmentally friendly technologies are directions under which subsidies and soft loans are provided. The maximum subsidy level for entity that executes commercial activity is 199.8 thousand EUR, and for entity that executes non-commercial activity, it is 1.5 million EUR. In any case, subsidy cannot exceed 80% of total eligible cost. In 2012, invitation to submit proposals under measure the "RES utilization in individual buildings, which were built in accordance to technical building norms applied till 1993" was published. Total financing resources that could be allocated were 260.66 thousand EUR. Subsidy could cover 30% of total eligible cost. Up to April 2013, 72 thousand EUR were allocated in a form of subsidy for RES project implementation under the measure.

Banks and credit unions also provide financing to RES electricity generators in a form of a loan. The first bank that was interested in RES projects was Swedbank. During 2005–2009, Swedbank invested more than 130 million EUR in environmentally friendly technologies, mainly, wind. The interest of bank in financing wind parks increased significantly in 2009, when subsidy in a form of feed-in tariff increased from 5.8 EUR ct/kWh to 8.7 EUR ct/kWh. The largest Lithuanian bank, SEB bank, also provides loans to wind park development. In 2009, SEB bank provided a loan of 7.24 million EUR for construction of a wind park, and later, its loan increased to 16.25 million EUR (covered 75% of total project value) for the construction of a new wind park of installed capacity of 13.8 MW. Besides, 42 million EUR loan in equal shares with Finish bank Pohjola Bank Plc for construction of the largest wind farm (of 39.1 MW) in Lithuania was provided. SEB bank also funded attractive biomass technologies; however, only since 2009, when subsidy in a form of feed-in tariff increased from 7.0 EUR ct/kWh to 8.7 EUR ct/kWh. In 2009, SEB bank provided 6.66 million EUR loans for construction of biomass fired CHP.

Banks interest in RES sector highly increased, when the Law on Renewable Energy Sources was implemented, new and high feed-in tariffs were approved. Since high and long-term feed-in tariff for solar electricity was approved, the demand for such type of plants drastically increased. The interest of banks in the solar market also increased. As a result, Medicina Bank decided to finance two solar PV projects, the total installed capacity of which is 1 MW. Total project value is 2 million EUR. Value of the first project, which has already been implemented, was 0.98 million EUR. Medicina Bank provided a loan of 0.75 million EUR with the loan repayment period of five years (Medicina bank, 2012).

Heat Sector

There is no direct subsidy in a form of feed-in tariff for heat produced at the plants utilizing RES.

Environmental Investment Fund. However, subsidy and soft loans for RES heat technologies are provided under the EIF programme. In 2010, 3.79 million EUR were invested in biomass technologies used to generate heat through EIF programme. Investment volume reduced to 1.24 million EUR in 2011. Investment through CCPS was also provided. With reference to data of 2012, 5.7 million of subsidies were offered to finance biomass projects.

EU Structural Funds. Subsidies to heat sector are also provided from EU Structural Funds for a period of 2007–2013. Lithuania implements Cohesion promotion action programme. In the framework of this programme, subsidies are provided under the measure "RES utilization for energy production." At the end of 2012, beneficiaries asked for 164 million EUR, but 29 financing agreements were made, the total value of which is 61.9 million EUR. The subsidy level varies in a range of 28.96 thousand EUR – 5.2 million EUR. Intensity is 50%. The remaining financial resources are covered by the private investors and using individual financial resources. No governmental assistance is foreseen for the implementation of the measure. There are examples, when biomass related projects are co-financed using various financing channels; however, financial instruments are typical –subsidy and loan. In 2010, in Siauliai city, building of biomass based combined cycle heat and power plant had begun. The support from EU Structural Funds was 5.2 million EUR. Heat generating company used 1.0 million EUR of its own financial resources for the project. The other financial resources were received from European Investment Bank (10.89 million EUR) in a form of a loan. SEB bank and Swedbank (13.61 million EUR) provided a syndicated loan.

Tax measures do not make incentives to use RES in heat sector, except exercise tax, which considers that biomass is not taxable.

The analysis of financing channels and instruments in Lithuanian heating sector disclosed that biomass is recognized as a type of RES, which has possibilities to penetrate Lithuanian heating market, which currently is based on combustion of natural gas. However, thus far there is no information about possibilities of other RES technologies to enter heating market. In this way, this paper will fulfil the gap and will reveal the place (competitiveness) of solar collectors in portfolio of energy generation technologies.

Integration of renewable energy projects (such as solar collectors) into district heating systems may create a positive external effect on the entire society concerning environmental and other regional development goals.

Consumers' need for hot water or space heating is satisfied without burning natural gas or biomass; therefore, the pollution is avoided. Moreover, import of natural gas from Russia is reduced, and the forest conservation increases.

Promotion of the competition and support schemes that enable competitiveness are the measures that could give greater opportunities for the use of solar and geothermal energy in the district heating system. Energy, which is produced by consumers, can participate in the competitiveness process if this issue is treated on the state level, i.e., minimize investments and costs in the macroeconomic context. In this regard short study of the financial system and opportunities for improvement is given.

Heat producers that use RES technologies have no direct subsidy, such as feed-in tariff. However, subsidies for investment and soft loans for RES heat technologies are provided by the Lithuanian Environmental Investment Fund (LEIF). In 2012, 5.7 million EUR of subsidies were offered from LEIF to finance biomass projects. Subsidies decreased to 3.1 million EUR in last year's financing period for biomass projects.

Subsidies for the heating sector were also provided from EU Structural Funds for the period of 2007–2013. Lithuania has implemented the Promotion of Cohesion programme. Subsidies under the measure "Utilization of RES for the production of energy" were provided under the mentioned programme. Beneficiaries were asking for 164 million euro in 2012, but finally, only 29 financial agreements for total value of 61.9 million euro were signed. The amount of subsidy varied from 29 thousand euro to 5 million euro with intensity of fifty per cent. The remaining amount of financial resources must be covered by the private investors using their own financial resources or loans. Governmental assistance is not foreseen for the implementation of the measure. Several examples of co-financed biomass related projects by various sources of financing exist; however, typical financial instruments, such as subsidy and loan, dominate. The fourth largest city of Lithuania Siauliai in 2010 started building a combined cycle heat and power plant, fuelled by biomass. The support from EU Structural Funds was 5.2 million euro. Heating company used about a quarter million euro of private financial sources for the project. Moreover, additional financial sources of 10 million euro were received from European Investment Bank as a loan. Scandinavian banks SEB bank and Swedbank provided a syndicated loan of more than 13.6 million euro.

Tax incentives for the utilization of RES in the heating sector are not provided, apart from the excise tax that is an exception for biomass.

European countries use various financial sources and different incentive mechanisms to support energy from RES. However, traditional financing sources and incentives are used in Lithuanian RES sector because of the lack of knowledge, shortage of experience and unstable regulation of RES sector.

Possibilities to use innovative financial sources and incentives for the wider finance of RES sector in Lithuania are presented in next section concerning JESSICA initiative.

JEREMIE (Joint European Resources for Micro to Medium Enterprises) Holding Fund was established at the end of 2008 in Lithuania after Ministries of Finance and Economy and European Investment Fund reached an agreement for establishment of JEREMIE Holding Fund. Total sum of 210 million euro was directed to this fund. Financing is provided to small and medium enterprises via financial intermediaries in a form of venture capital and guarantees. JEREMIE can support improved access to finance in urban areas.

An overview of different sources of finance for district heating suppliers and consumers revealed that market is ready to finance implementation of RES alternatives.

The risk of biomass and other fuel prices, and lower demand of heating after renovation of multifamily buildings should be taken into account in further researches.

Moreover, further researches should take deeper analysis of the macroeconomic effect that arises from the interaction between district heating producers and consumers.

Monopoly of district heating exists due to specific of the infrastructure. The detrimental trend is to solve complex tasks by one-off measures. Uniform programmes are needed to relate both spheres of production and supply. Nearly all support measures are concentrated on the production side in Lithuania, while the infrastructure could be used for the consumers' efforts to reduce the volume of consumption and to produce heat and hot water by the use of solar and geothermal energy without disconnecting from district heating.

Reforms in heat sector, Renovation (Modernisation) Programme of multifamily buildings, and implementation of Law on Renewable Energy are needed in order to make purposeful, legal scenario system; then district heating suppliers and consumers could interact, and investment decisions could be synchronized in unified investment scenario.

According to the Heat Supply and Consumption Rules in Lithuania, multi apartment building or one apartment could disconnect from district heating system and use another energy source. However, the Rules exclude the cases when consumers install heat production facilities of RES without disconnecting from the district heating. Outstanding question remains how to cover fixed costs of heat supplier, when such consumers use significantly less amount of heat than in case without RES facilities. Nevertheless, heating system must be maintained at appropriate technical conditions so that at any time, these consumers could have supply of heating.

4.2.2. Promotion of Wind Energy Development

On February 3, 2012 National Commission for Energy Control and Prices (NCC) approved the new electricity, which is produced by using RES, feed-in tariffs (Table 4.7).

Table 4.7. Feed-in tariffs and maximum tariffs for electricity, produced by wind power plants (Source: NCC)

The power plants by technology type	The installed capacity (IC), kW	Tariff, LT/kWh excluding VAT	
		2011	2012
Wind power plants	IC < 30 Feed-in tariff	0.09	0.11
	30 < IC < 350 Maximum tariff		0.10
	IC > 350 Maximum tariff		0.08

The Commission recalculated and on September 27, 2012 approved fixed rates for producers that use RES. The Commission considered the development of production of electricity from various RES, also with the regard to the actual amount of electricity produced during the previous calendar year and the sum of installed capacity of operating plants as well as the sum of intended to be installed capacity of plants under construction. Tariffs for electricity that is generated at wind power plants remained unchanged.

On February 28, 2013, these tariffs were recalculated again (Table 4.8). The reduction was not only for tariffs, but also for power limit below which feed-in tariff is applied (10Kw instead of 30 kW). From February 1, 2013, the amendments of the Law of RES entered into force, thus the definition of a small wind power plant changed: small power plant capacity is limited to 10 kW. Therefore, small wind power plants to be considered in terms with the Law have to be no more than 10 kW of installed capacity.

Table 4.8. Feed-in tariffs of surplus electricity and maximum tariffs for producers that intend to participate in the auctions, EUR/kWh (Source: NCC)

Wind power plants			
The installed capacity (IC), kW	Before April 1, 2013	From April 1, 2013	The installed capacity (IC), kW
IC < 30	0.11	0.10	IC < 10
30< IC < 350	0.10	0.09	10 < IC < 350
IC > 350	0.08	0.08	IC > 350

Commission approved new tariffs on electricity produced from RES that will be in force from April 1, 2013 till June 30, 2013. Taking into account amendments of the Law on Renewable Energy that entered into force on February 1, 2013, the Commission is obliged to recalculate tariffs for electricity produced from RES no more than 4 times a year.

The development of wind power plants is encouraged by support for investments. Finances for measures, implementation of which leads to reduction of quantitatively calculated emissions of greenhouse gases, were provided by order of minister of Environment of the Republic of Lithuania in the 2012 estimate of Special Climate Change Programme.

Support for installation of small wind power plants is provided by the three measures (Table 4.9).

It should be noted that the support is given only for the following activities:

- installation of wind energy sources for individual use (not for sale);
- electrical grid adaptation;
- other specific installation equipment that is needed for wind power plant project.

According to Lithuanian Environmental Investment Fund (LEIF), projects that received funding under the second measure of Special Climate Change Programme estimate of the use of 2012 funds (part 1, paragraph 1.2.1) “The use of renewable energy sources (solar, wind, geothermal, biofuels, etc.) in individual residential buildings, which were built according to technical specifications of construction valid until 1993" had no projects of wind power plants in the 2012 list.

Table 4.9. Special Climate Change Programme measures, which provides support for installation of wind power plants in 2012 (Source: Lithuanian Environmental Investment Fund)

Name of the measure	Amount, million EUR
The use of renewable energy sources (solar, wind, geothermal, biofuels, etc.) in public buildings and business enterprises	30*
The use of renewable energy sources (solar, wind, geothermal, biofuels, etc.) in individual residential buildings, which were built according to technical specifications of construction valid until 1993	4.5**
The use of renewable energy sources (solar, wind, geothermal, biofuels, etc.) in multifamily buildings, which correspond to the D energy efficiency class of the house and were built according to technical specifications of construction valid until 1993	1.5***

* The amount of subsidy is 50% of the total eligible costs of the project for applicants from business enterprises and 100% of the total eligible costs of the project for applicants from non-business enterprises

** The amount of subsidy is up to 30% of the total eligible costs of the project.

*** The amount of subsidy is 15% of the total eligible costs of the project, and subsidy is provided if the applicant did not receive or cannot receive State support for the implementation of these measures according to the Programme for the Renovation (Modernization) of Multifamily Buildings or corresponding municipal programmes

Table 4.10. Maximum amount of subsidy for wind power plants (Source: Lithuanian Environmental Investment Fund)

Technology/equipment	Unit	The maximum amount of subsidy EUR/unit
Wind power plant: horizontal axis without batteries	kW	3071
Wind power plant: horizontal axis with batteries		2764
Wind power plant: vertical axis without batteries		2457
Wind power plant: vertical axis with batteries		2212

Recently, LR Ministry of Environment approved financing trends of 2013, including granted amount of funding (around 0.6 million euro) for the use of renewable energy sources (solar, wind, geothermal, biofuels, etc.) in individual residential buildings.

In order to ensure the efficient use, maintenance, renovation, and modernization of the existing houses, also rational use of energy resources, the Government of the Republic of Lithuania in decree No. 1213 (Official Gazette, 2004, No. 143-5232; 2009, No. 112-4776) from September 23, 2004 adopted Programme for the Renovation (Modernization) of Multifamily Buildings. This programme aims at apartment owners' promotion for the renovation (modernisation) of multifamily buildings in order to improve the quality of life, use energy sources rationally, and to reduce budged spending on compensation for the residential heating. Installation of alternative energy sources (solar, wind, etc.) is funded by the implementation of this multifamily building modernization programme. The investment is funded according to European Commission initiative JESSICA.

4.2.3. Possible RES Financing Channels and Instruments

The worldwide experience suggests that various financial channels and instruments are used to support energy from RES. Lack of knowledge, shortage of experience and low stability of regulation of RES sector in Lithuania dictate the mode that traditional RES sector financing channels and instruments are used in Lithuania (EU structural funds, 2013). However, this section of the monograph will disclose innovative financial channels and instruments that could be more widely used to finance RES sector in Lithuania.

JESSICA Initiative

JESSICA initiative is a Joint European Support for Sustainable Investment in City Areas. It is an initiative of the European Commission developed in co-operation with the European Investment Bank (further in the text EIB) and the Council of Europe Development Bank (CEB). It supports sustainable urban development and regeneration through financial engineering instruments such as stock capital, loans and warranties. Since the beginning of 2008, Lithuanian Ministry of Finance together with EIB analysed JESSICA implementation possibilities in Lithuania. It was decided that JESSICA funds would be allocated to the sector of old multifamily houses, which were built before

1993. It was agreed that funds must be directed to implementation of energy efficiency measures. For this purpose, JESSICA controlling fund was established. 227 million EUR (127 million EUR came through EU structural assistance, and 100 million EUR from the state) were directed to implementation of this measure, which aims at improving financing conditions of multifamily houses, higher education schools and student dormitories. EIB selected three banks (namely, Swedbank, SEB bank and Siauliu bank) in Lithuania, which finance projects. Banks finance up to 100% of construction costs. Soft loans are provided for 20 years with 3% fixed interest rate. JESSICA controlling fund foresees support of various activities related to repair and reconstruction of the building. It is also favourable to RES sector, since it provides assistance in a form of soft loans for installation of RES technologies in buildings. This is an important channel for financing RES sector in Lithuania, and it should be more widely applied.

JEREMIE Initiative

JEREMIE controlling fund was established at the end of 2008 in Lithuania after Ministries of Finance and Economy and European Investment Fund contacted an agreement for establishment of JEREMIE controlling fund. 210 million EUR were directed to this fund. The peculiarity of the fund is this: it provides risk capital to SME through several risk capital funds:

– *LitCapital I* (JSC "LitCapital Asset Management" is its manager) was established in 2010 as a development capital fund, which invests in the business development of Lithuanian private micro, small and medium-sized enterprises. Fund is oriented at several branches of the economy: namely, IT, services (B2B, B2C), manufacture of commodities and bioenergy sectors. Fund investments are long-run and are made in a form of purchasing stocks of the company. Mainly, fund invests 0.6–3.0 million EUR for 3–6 years in fast growing companies. After investment period ends, it sells stocks. Currently, the portfolio of the LitCapital consists of seven companies, one of which is directly related to RES sector. This is bioenergy JSC "Lignoterma." JSC "Lignoterma" signed an agreement for 0.87 million EUR (planned total project value 6.4 million EUR). These financial resources will be used to construct 26 MW biomass plant in Panevezys city. Plant will produce heat energy that will be supplied to one of the biggest manufacturing companies in Panevezys, SC "Amilina."

- *Lithuania SME Fund* was established in 2010 by Baltcap fund. It will invest in the promising Lithuanian small and medium-sized enterprises. Currently, Baltcap is the largest private and risk capital fund in the Baltic States. Lithuania SME Fund will invest 20 million EUR in 7–15 Lithuanian companies, which will be selected by Balcap.
- *Business Angels Fund I* invests in the promising Lithuanian small and medium-sized enterprises, turnover of which reaches 4.3 million EUR a year, sales oriented towards export markets, company functions 1.5–2 years. Fund invests only together with business angel. Fund together with business angel seeks to acquire 50% of stocks of the company. JSC "Baltic Sun Energy" has been already included in the portfolio of the fund. Fund and business angel financed solar power plant.

INVEGA Fund

INVEGA fund was established in 2009 on the basis of JEREMIE initiative. Fund activities are based on a lending principle, which means that fund provides selected banks with the financial resources. Later, banks lend these financial resources on their own risks to SME. Fund does not accept risks related to lending, as well it does not cover loan administration costs that are sustained by banks. Financial resources are allocated to SME in a form of soft loan, conditions of which are agreed upon with a bank and are a result of negotiations. INVEGA fund provides up to 100% capital. Bank must pay 3–4% interest rate to the fund for its financial resources. INVEGA fund manages the following measures: small credit provision-2 stage, open credit fund, shared risk loans, portfolio guarantees. These measures are favourable to RES business financing. In a framework of the measure "Small credit provision-2," RES company could receive a loan up to 100 thousand EUR. This loan is provided to finance investments and working capital. When banks provide RES companies with loans, they cannot ask for a mark-up that is not higher than 2.2%. Considering the fact that basic interest rates are very low, this seems an attractive option to RES business to borrow money during the slow economic recovery period. Measure "Open credit fund" could be attractive to RES companies that require finance for investment and working capital up to 435 thousand EUR. The experience of Lithuanian companies disclosed that during 2009–2013 (first quarter), more than half of credits were supplied to finance working capital. These credits were provided with higher interest rate (3 month EURIBOR or VILIBOR + 0.1% + bank mark-up, which cannot be higher than 3%) than under the measure "Small credit provision-2." Large

RES project could be financed using measure "Shared risk loans." Under this measure, RES projects could receive loans up to 4.8 million EUR. RES companies could also use measure "Portfolio guarantees." This measure is dedicated to promote SME lending by guarantying 80% of loan. In this case, the barriers to RES company to receive finance, when there is a shortage of collateral, would be reduced.

Table 4.11. Comparison of Practica funds (Practica, ESparama)

Criteria	Early stage risk capital fund	Risk capital fund
The main investment stage	Early stage "seed, start-up" business financing.	Development financing (working capital, investments.).
Object for investment	Initial business ideas, projects, micro, small and medium-sized enterprises, which develop new business concepts.	Existing micro, small, medium-sized enterprises.
Geography:	Enterprises established and functioning in Lithuania.	
The main form of investment	Ownership (shares) and semi-ownership (convertible debts) capital.	
Planned shares of ownership	Small and large portfolios of shares or semi-ownership financing.	
Size of investment	3,000–200,000 EUR per project.	200,000–3 million EUR per project.
Investment period	2–5 years.	2–5 years.
Business sectors	Various business sectors, priority to IT and innovative business.	Various business sectors.
The main selection criteria	strong team; quality of business idea; large market or clear niche of the market; possibility to test the idea with relatively small investment.	clear business model; strong team of leaders; priority to strong business with good history, share of domestic market, positive and stable cash flow, high capital return.
Participation in management	Active investor who contributes to creation of high value added and is on the board.	
Exit from activities	Selling to strategic investor or business leaders, as well bank refinancing.	

INVEGA fund also offers "Innovation development" measure. Measure aims at promoting transmission of ideas generated in scientific institutions to business. In this case, RES companies could practice the most innovative business development ideas created by scientific institutions. Risk capital is provided within the framework of this measure. Measure is implemented through Practica Capital company, which manages two funds: Practica Seed Capital fund and Practica Venture Capital fund. Comparison of funds is provided in Table 4.11.

Thus, results of analysis suggest that JEREMIE initiative should be attractive to RES companies, since innovative measures and good financing engineering instruments are provided under the framework of this initiative. It provides real opportunities for RES companies to solve business financing issues.

Chapter 5

LONG-TERM MEASURES FOR JUSTIFICATION OF IMPROVEMENT OF RES SUPPLY AND DEMAND SUPPORT MEASURES

5.1. KNOWLEDGE BASED MEASURES TO PROMOTE THE USE OF RES

Enhanced financing of environmentally friendly technologies by using renewable energy sources (RES) is necessary seeking to overcome energy market failures and barriers. EU Structural Funds are very effective financial channels from this point of view. These funds can be used effectively by new member states to overcome a set of market failures by financing innovations and diffusion of new environmentally friendly technologies.

Demand of RES technologies is not adequately restricted by slow demand rates. Just in case of huge demand, progressive technologies may have economic background for a successful development. The main problem is the increasing public interest, which may be expressed through continuous supply of energy resources in future, solving of social problems, environmental solutions and also inability of private sector to invest in current conditions. However, the awareness of energy consumers usually contains just scraps and fragments in comparison to the full effective knowledge, which forms demand based on public benefit.

Towns and regional energy programmes are a prospective organizational and financing form of integration of technologies using Structural funds. However, integration of RES may be more successful after reliable methodological assessment of positive effect for solving social, economic, development problems in regional development context.

Scientific and practical problem is being solved while evaluating links between energy development and economic growth with providing possibilities to solve social and security problems.

Organizational forms are extremely important for RES implementation on a wider scale. These are various regional or municipal programmes, which include RES projects to solve energy supply and environmental (e.g., biogas plants in stock-raising farms and municipal waste incineration plants, etc.) and social problems in rural areas as well as many other functions in the context of sustainable development.

The exchange of practices, based on sustainable energy aspects, is the key to promotion of the development of projects at local level.

Integration of energy projects into regional development process may create external positive effect concerning environmental and other regional development goals.

Sustainable energy investment projects, characterised by a positive local environmental development impact, can be brought up to the level of implementation using contribution not only from Structural Funds but also from other regional public resources.

The need for formation of single energy policy with RES economic support system as one of compounds is notified for several years already. It is impossible without single attitude to the perspectives of the use of separate types of fuels and sorts of energy in making strategic decisions as well as in implementing them (Dusonchet and Telaretti, 2010b).

It is necessary to separate two main support groups concerning RES:

a) Energy producers and suppliers, which must integrate RES into operating energy system to compete with main types of fuel – natural gas, oil (heavy oil fuel), coal;
b) Energy consumers to whom RES without additional support and establishment of infrastructure cannot afford most demand defining factors – acceptable price and security of supply.

However, the use of RES is related to additional public benefit, which is not evaluated in individual business decision-making.

RES accumulates the following main qualities:

- RES means inexhaustibility of their use, which means possibility to supply future generations with energy sources;

– Implementation of RES means that technological progress is directed towards harmony of human activity with natural processes;

The scientific problem lies in definition of the economic background for establishment of a uniform system of economic and legal tools, measures and mechanisms. It is necessary to analyse economic support forms for creation and mastering of RES technologies to make them competitive with traditional fuel technologies and make them sensible according to the actual purpose, need and possible effect. All support schemes must form uniform, a systematic and blameless economic whole. In principle meaning, this support scheme must make RES compatibility conditions the same, but not protect them against other sorts of energy and technologies.

Research in economic theory is very intensive. The main trends of investigation related to RES economic support policy are as follows:

a) Analysis of various financing related risks and barriers, which suspend a wider development of the RES sector.
b) Assessment of external costs and including them in costs price, caused by use of traditional energy sources – oil and oil products, natural gas, coal.
c) Analysis of the problem of exhaustibility of resources, which participate in economic process and reflection of this feature in economic estimations.
d) Investigation of issues, related to the use of developed RES technologies, e.g., hybrid wind-hydrogen plants, tri-generation and many others.
e) Creation of organizational management forms, which permit assessment and financing on a wider scale.
f) Creation of innovative financial tools and mechanisms, enabling RES financing.

Economic and financial support of RES certainly does not lie with the intrinsic goals being pursued by competitive market. For example, implementation of RES may increase the security of energy supply for a company, district or even region. Projects carried out for RES implementation provide improved comfort and have manifold effects on new jobs, dispersed across the community both socially and spatially and involving small and medium size enterprises.

Development of renewable energy sources is in fact a part of implementation of climate change policies, improving energy efficiency and ensuring security of energy supply (Renewable Energy and Energy Efficiency Partnership, 2012). Most important is the recognition that they all are necessary elements of a sustainable energy, future that will not be delivered by market forces alone. Many of the benefits of RES are the public benefits that accrue to everyone. For example, those who choose RES, reduce pollution for everyone and provide an environmental benefit to the entire society. However, a customer who is willing to pay more for electricity from RES still has to breathe the same air as the neighbour, who might choose not to pay more.

Employment, fuel diversity, price stability, and other indirect economic benefits of RES are provided for the society as well. For example, for a large industrial customer, it may make more sense to risk moving to another region in response to increases in fuel prices rather than pay more for RES to stabilize regional prices. While this strategy may benefit the individual firm, it is likely to hurt the region's long-term economic competitiveness. In the same way, firms that can pass on increases in energy costs to customers may also lack an incentive to diversify fuel sources, even though investment in RES would stabilize prices over the longer term.

However, sustainable growth in this area can only be achieved through a regulatory framework, which complies with the rules of a competitive market. Maintaining that support schemes for RES therefore have to be market based, capital efficient and designed to avoid market distortions. EURELECTRIC calls for harmonization of RES support schemes.

Creation of competitive energy market is very favourable for the increase in economic growth; however, security of energy supply, mitigation of environmental impact and energy affordability are the targets of sustainable energy development, which can be achieved only by implementation of wise Government energy policy. Sustainable development needs organizing institutions, actors, support measures, etc. The most important thing is to understand that the main essence of all EU directives, policies and measures is to overcome energy market failures: regulation of monopolies in energy sector, dealing with energy externalities, energy affordability, asymmetry of information and energy security. Implementation of RES technologies is a key element of sustainable development strategy.

The problem certainly does not lie with the intrinsic goals being pursued as a part of the completion of a competitive market. "The problem lies with current inconsistencies, lack of market integration and frequent absence of a least-cost approach. A patchwork of policies, burdens and new regulations, keeps building up without sufficient coherence, balance or properly thought-out economic assessment. As a result, the essential aims of regulatory consistency, avoidance of market distortions and use of least cost solutions are far from being met" (EURELECTRIC, 2004).

It is clear that the main stakeholders in this process are market forces. However, state, municipalities, nongovernmental organizations, consulting firms may play a considerable role as a connecting, integrating and promoting chain between the financing organizations and energy consumers (Vos and Savin, 2012).

EU Structural Funds have a huge role in this field; however, their distribution needs more rationality and more orientation towards sustainable development.

The EU policy for sustainable energy represents ambitious targets and requires huge investments. The gap between objectives and available resources is the main weak point, which undermines the possible fulfilment of the EU policy objectives (EC, 2014). A further weak point concerns the fact that the size and the impact of integration and implementation of EU energy policies and related programmes into the regional and local environment are not clear. A possible solution to this problem can come from the involvement of the regional development resources and the Structural Funds in the energy policy related objectives and programmes (Antinucci et al., 2002). The sustainable energy objectives may be reasonably integrated in the regional development strategy in an overall vision of a sustainable development for the region.

It is clear that market forces are the main stakeholders in this process. However, municipalities may play considerable role as a connecting, integrating and promoting chain between the financing organizations and energy consumers. "Formulation of national sustainable energy strategies must emerge from interactive and inclusive processes of social dialogue and reflection. The processes are guided by orientation knowledge of justification arguments used by different stakeholders" (Laes and Verbruggen, 2010).

Talking about perspectives of the use of solar energy, it is evident that the main market is on a user's side. The most important problem, determining the slow absorption process of solar energy on the part of the consumer, is the lack of knowledge and organisation, deterrent amount investments, and especially the differences between energy suppliers and users in the heating sector.

It has to be noted that technology development is noticeably one step ahead the general level of society education and its susceptibility to innovation, in economic terms, the demand of new technologies, innovative forms of financing and forms of managerial organization. For example, the sustainable development of the heating sector is an integral part of smart grids that are based on the idea of developing information and communication technologies. It is the reorganization of the electricity and heating system, based on technologies and innovations, the main idea of which is to develop sustainable cooperation between producer and consumer, giving users more opportunities to be in the electricity and heating market. Development of smart grids would promote the use of RES and also promote the liberalization not only of the electricity, but also of the heating market. Reorganized system of heating market would stimulate the efficiency of consumption and production of electricity and heating, would decrease the negative influence on the environment, would motivate the creation of new workplaces and would reduce the prices of energy.

A new educational conception is needed, directed at the development of mentality of the society for sustainable development. Moreover, the concept of sustainable development determines the boundaries and absolute limits, and restrictions placed on the influence of current technologies on environmental sources and the necessity to absorb the effects of human activity (Klevas et al., 2009).

The task of RES evaluation makes one deepen one's knowledge and answer to more than one question on the economic theory. In the presence of global risks of climate changes a question arises, what is the essence of technical progress in the presence of exhausting energy sources and how the economic theory and policies have to be transformed? Because of the fluctuation of natural gas prices, the development of economy becomes more and more problematic and difficult to forecast, and the combustion of energy sources is certainly accepted as an incentive of the global climate changes. Therefore, the necessity of the use of RES, as widely as possible, is undoubted.

In this context, the solidity of a knowledge-based economy concept emerges. One of the main roles of science is to give information, which enables the formation of the development policy in a decision-making process. It is necessary to essentially improve scientific perception, based on scientific researches, and to raise the scientific competence of a country.

The concept of knowledge-based economy noticeably expands the general concept of development and adequately raises new goals and assumptions that are defined in Figure 5.1.

Only complex knowledge and appropriate conveyance of it and the system of education can form an economic base to successful development of progressive technologies. However, energy users' knowledge about the benefits of RES is only fragmented compared to the entirety of knowledge that creates a demand based on the social utility. Therefore, investments in the dissemination of knowledge can be approached as components promoting technical progress. Consequently, the greatest benefit can be provided by knowledge, formation of reliable and practically proved knowledge, the dispersion of knowledge and promotion of technology design.

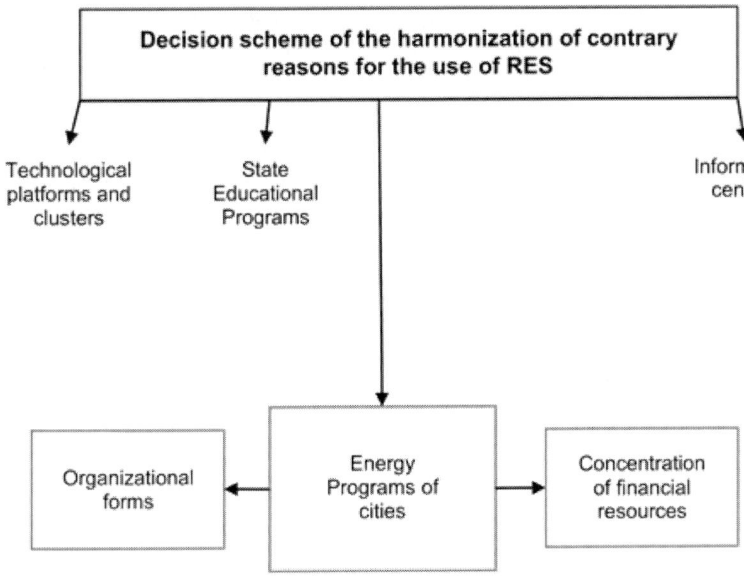

Figure 5.1. Decision scheme of adjusting different motives in regard to RES.

5.2. Implementation of National Programmes via Local Sustainable Energy Development Plans

The energy use in Lithuania is considered a basic driving force that impacts environment. It is a major contributor to environmental problems of global concern, such as climate change, acidification and urban air pollution (Strategy for Development of Renewable Energy, 2010). Energy is crucial for economic and social development, quality of life and consumption level. According to the concept of sustainable development, it is necessary to incorporate goals of sustainable development into all sectored policies. However, some different documents have already been adopted: National energy strategy, National sustainable development programme, National regional development programme.

The main long-term planning document is Long-term Lithuanian Economy Development Strategy approved in 2002. It comprises fifteen branch strategies. The main principles of sustainable development are integrated in these strategies. Some of these branch strategies are directly aimed at interaction between sectors (the factors of social development and economic factors of employment, economic factors of environmental protection, tourism development, etc.). Despite a great integrity of the certain strategies, there is a lack of clear relations between the aforementioned fifteen strategies. In order to solve this problem, the National Strategy of Sustainable Development was adopted in 2003. This strategy includes six branches of economy (transport, industry, energy, agriculture, household, tourism), four environmental sectors (air, water, biodiversity and waste), four main social aspects (employment, poverty and health, education, cultural identity) and regional development issues. All these economic, social, environmental and regional development issues are presented in close integrity. Sustainable development indicators for economic, social and regional development and state of environment are selected in the strategy for monitoring sustainable development; however, this system of indicators was not applied to the analysis of trends, and only some targets of sustainable development were set using these indicators (Ciegis & Streimikiene, 2005). Approach to integration of sustainable development indicators into national planning systems should be used at a greater extent following these steps:

- To set targets of sustainable energy development;
- To analyse trends and interlink ages of selected indicators of sustainable development;
- To formulate response actions based on performed analysis;
- To monitor the progress achieved towards set targets;
- To evaluate efficiency of implemented policies.
- Sustainable development strategy for 2002-2020 establishes the main targets relevant to sustainable energy:
- Economic growth balanced between economic branches and regions;
- Minimization of social and economic differences between regions and within regions;
- Minimization of environmental impact from the main sectors;
- More efficient use of natural resources and waste management;

- Minimization of impact on human health;
- Climate change mitigation.

The main targets relevant to sustainable energy development expressed in indicators:

1. To reduce energy intensity of GDP by 50% up to 2020;
2. To reduce CO_2 and other pollutant emissions per GDP unit by 50% by 2020;
3. To reach the share of renewable – 12% in TPES by 2010 and 15% by 2020;
4. To reach the share of CHP – 35% of total electricity production by 2020;
5. To reach the share of renewable – 10% in total electricity production by 2010;
6. To reach the share of bio fuels not less than 15% in transport fuels by 2020.

The main objectives of the National Regional Development programme are as follows:

1. To promote economic restructuring and modernization in regions;
2. To promote sustainable development of regions;
3. To reduce social-economic differences among regions;
4. To reduce unemployment;
5. To develop rural areas by converting and restructuring their economies, reforming agricultural production, manufacturing industry and marketing structures, to promote establishment of new businesses and services in rural areas.

All these targets should be implemented on the regional levels. The urban strategic development plans of cities should integrate sustainable development targets established by National sustainable development strategy. For example among these priorities, the main goals and objectives were defined which fit well within sustainable development targets established for Lithuania in Sustainable development strategy. Using the best practice case studies database, prepared during RUSE activities, the following projects can be implemented in Vilnius matching the priorities, goals and objectives of Vilnius strategic plan and the targets of Sustainable development strategy. Using the data on the good practice case studies from the database developed by RUSE project, one can conduct a theoretical experiment: integrate the complex of good practice case studies (GPCS) in Vilnius city strategic development plan taking into account the priorities of Lithuanian National sustainable development strategy that also include priorities of Sustainable energy development strategy (Klevas et al., 2009).

With regard to energy, the main methodological problem is integration of energy projects into regional development procedures, so that energy elements may compile integral uniformity in terms of regional goals. The main problem is formation of single economic state policy towards RES. It is not enough to just adapt various directives, regulations of international obligations. It is necessary and possible, as can be seen from foreign experience, to perform this on a single background in most beneficiary conditions for the sake of national economy and environment. It is necessary to perform the research, which could be a scientific background for making economic decisions (Klevas et al., 2010). The sustainable energy development guidelines can be elaborated based on implementation of sustainable energy development strategy in the framework of city strategic development plans (Figure 5.2).

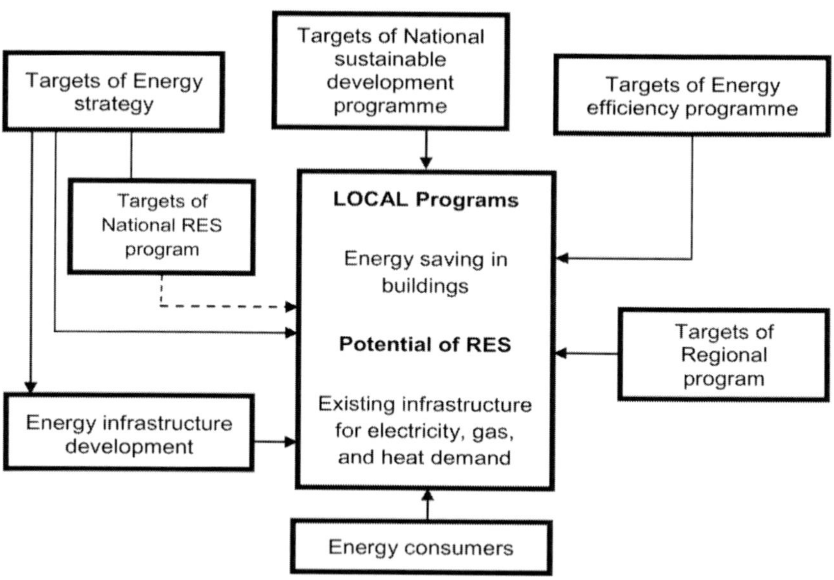

Figure 5.2. Interconnections between national and local energy programmes concerning RES.

The underlying principle for implementation of energy strategy, regional programme, climate policy implementation is therefore in that municipalities decide for themselves the topics on which they will focus their policy. They know better than anyone where the best chances of success lie. They can build on previously developed environmental and other measures. Moreover, they understand how to make the policy fit local circumstances and needs. For example, this means that municipalities in the densely populated urban agglomeration will be more likely to focus attention on energy-saving measures in the renovation of old housing stocks, whereas municipalities in the wind-rich coastal provinces will see more opportunities in wind energy.

Energy saving, energy efficiency measures and RES most often have additional positive effect, which complies with regional development objectives. For example:

- RES, being indigenous sources of energy improve the security of energy supply and diversity in the fuel mix.
- RES have considerable advantages for regions where power and heat supply costs are considerably higher than average costs in the Republic. Thus, use of RES electricity and heat in small isolated systems can also help to avoid or delay expensive extensions to the grid.
- Some RES are a labour intensive form of industry that creates jobs especially at location sites in rural areas.

The listed-above advantages ought to be included into evaluation process in feasibility studies. However, most often, decisions made by municipal and regional administrations, energy distribution companies, power supply companies are based on feasibility studies, dealing with particular investments either in energy conversion or in end-use system, or in the system of energy source production. Decisions based on such partial feasibility studies may not result in least cost solutions for the system as a whole (Klevas, 2010).

- Prices of all energy resources must be based on social costs with the assumption that it is methodologically possible to justify all aspects in the use of fuel and energy: security, environmental, social and renewability;
- RES benefit can be assessed according to real benefit now and in future, which is not fixed in investment efficiency with assumption that various sustainability criteria may have single denominator on single methodological background;

- Justification of RES support measures is based on comparison of fuel and energy resources used by energy system, on competitiveness while assessing various sustainable development aspects by a single criterion.
- RES competitiveness is evaluated not in general approach but in specific territorial environment till certain optimal level, where marginal costs and marginal benefit of all types of energy become equal.

With regard to definition of the guidelines for sustainable energy development in economic understanding, first, one should clearly evaluate current or necessary infrastructure, related to the used type of fuel.

Table 5.1 presents principal scheme for formation of region (town) energy programmes, which are considered a synthesis of energy saving, improvement of energy efficiency and implementation of renewable energy sources. However, with regard to attraction of investment and variety of financing possibilities, this general model is formed as a package of programmes, directing towards understanding that payback of investment projects should be based on realistic reduction of energy losses and energy saving.

The scheme of integration of energy projects into regional development policy could be illustrated as follows:

Table 5.1. Illustrative scheme of defining regional energy development scenario

Energy projects / Directions of regional policy	Renewable energy sources (biomass, wind, solar, etc.)	Energy savings measures	Improvement of energy supply efficiency	...
Solutions of social problems	a_1^1	a_1^2	a_1^3	...
Support for agriculture	a_2^1	a_2^2	a_2^3	...
Solutions of environmental problems	a_3^1	a_3^2	a_3^3	...
...	...	...	...	...
Total positive effect of implementation of energy projects	x_1	x_2	x_3	x_i

a_i^n – value expression of positive effect of solution of implementation of energy projects for social problems, support for agriculture, etc. in the region.

x_i – summarized total positive effect of implementation of energy projects.

The most difficult problem is to define a possible external positive effect (a) of energy projects in framework of regional policy directions. However, it is necessary for selection scenario of regional energy development, according to total positive effect of renewable energy projects, energy saving measures and improvement of energy supply efficiency. Based on such studies, regional authorities have to influence the decision-making by economic means. This could be done via fax exemptions and subsidies from budget. The government could remove financial obstacles to the made investments and make the sufficient purpose-oriented financial means available via the capital market and banks.

For the purpose of preparation and implementation of regions and towns energy programmes, functions for regional and municipal authorities should be formed and legally implemented.

There are many examples of effective practices: they perform RES market establishment and awareness raising functions. These examples should be investigated, scientifically generalized and proposals prepared for applying to Lithuania's conditions.

5.3. Assessment of Economic and Social Benefit from Using RES

5.3.1. Spatial Aspect for Assessment of Benefit from Using RES Technologies in Urban Areas

The stimulation of energy producers and consumers for the use of RES is one of the major goals of energy policy in Lithuania. Policies and measures that aim at enhancing the use of RES are mainly driven by EU policy. Unstable state energy policy and changes of incentive measures destabilize investments initiatives. The lack of both sides economic evaluation is the main challenge during the way to achieve target. An unreasonably huge role is given to biomass in the projections of the use of RES. Biomass is an energy resource, that is, the result of economic activity depends on continuity of economic activity. Meanwhile, the use of inexhaustible solar and geothermal energy potential is insufficient.

The overview of financing sources and instruments in Lithuanian heating sector revealed that biomass is recognized as RES type, which has the best opportunities to transform the district heating market in Lithuania, where natural gas is currently dominating. However, a lack of information about other RES technologies to enter district heating market is notified. Therefore, more diverse RES such as solar collectors and heat pumps should be analysed, especially comparing economy of RES on the supply and demand side of district heating.

The scientific problem is the evaluation of social utility of RES that can show the advantages, which are underestimated in investment decisions; for example, RES inexhaustibility and possibility to ensure sufficiency of energy resources for future generations at the same time. Utilization of some of the RES technologies, such as solar energy, also solves environmental issues. Therefore, they may be additionally financed from other sources. The main issue is the diversity of RES utilization opportunities and incentives. Moreover, it is a transfer of their economic interpretations on uniform rules and economic laws into a specified dimension, in this case, trying to reduce different opinions as much as possible.

The phenomenon of energy is RES that could change exhausted types of energy such as oil, natural gas and could be realized by few types of energy. This could be solar, wind, geothermal energy. On the other hand, social utility of RES differs from oil and natural gas, because RES guarantee for future generations the supply of energy resources, but this benefit does not have methodological evaluation. This is one of the main accents of the concept of sustainable energy development.

The external benefit analysis for small wind power plants, geothermal heat pump and solar collectors was performed to find their niche in urban and rural areas. The external effect indicators for different scenario cases are shown in Table 5.2.

The external benefit analysis for small wind plants was performed also to find their niche in urban and rural areas. The external effect indicators for all above-mentioned scenario cases are shown in Table5.2.

Table 5.2 shows comparison of external benefit using different type and scale of RES. Large scale (10000 m^2) solar collectors are the technology that generates heat mostly during summer period, when wind power plants and combination of geothermal heat pump and solar collectors in multifamily building gives effect during the whole year. The main indicators of external benefits are considered as the use of domestic capital, taxes paid to the state budget and created new job places. Combination of different types of RES may create large external effect and make a significant impact not only to individual consumer of RES, but to the society as well. Therefore, the benefit from using RES, especially in urban areas, should be evaluated during decision-making. If direct support were applied for RES or environmental technologies, the impact would depend on the volume of support and source of funding.

Table 5.2. External benefit of using solar, wind and geothermal energy technologies

Indicators	Wind power plants		Geothermal heat pump and solar collectors in multifamily building	Solar collectors in district heating system
	4 kW	10kW	126kW	7000kW
Produced, MWh	6.8	17.5	185	4500
Domestic capital, thousand EUR	8.02	21.08	25.01	1380.88
Social insurance, thousand EUR	0.81	1.51	9.27	119.23
Personal income taxes, thousand EUR	0.29	0.55	3.46	44.46
New job places (man months)	2.3	4.2	31.9	409.4

For example, if the direct assistance were financed from the state budget, the promotion measures would increase the budget deficit in addition to the above-mentioned positive aspects. It is therefore essential that the promotion is cost-effective: the benefits exceed the negative effects. Inexhaustible types of RES (wind, solar and geothermal energy) enable greater benefit than subsidies it may require while penetration into the market is growing.

5.3.2. Balancing of RES Support Forms as the Substance for Spatial Planning Method

Each of RES types has a very broad and distinctive use scale, so each renewable requires its own specific support (Wustenhagen and Menichetti, 2012). It was focused on the potential analysis of wind, solar and geothermal energy for rather narrow applications only. A survey of energy classification, financial, economic assistance forms, inventory, highlighting their sources of funding, performance indicators and the achievement of dynamics analysis should be performed. It is recommended to do this for all RES types.

Support measures for the use of RES often contradict market patterns due to short-term period goals, which are one of the main shortages in nearly all countries. These measures do not shape uniform technological programmes for long-term periods. The EU document (EC, 2014) on establishing support schemes for RES technologies define necessity to revise current RES support measures, which cause market faults, such as e.g., feed-in tariff increases supply; however, it also increases power and heat tariffs for all consumers, etc. This initiates constant changes of support tools, cause instability and thus distortion of investment environment.

Every renewable has its own specifics and wide and improving scale of technologies, which are becoming cheaper due to growing supply. Every renewable has different significance in terms of external benefit; thus, support schemes can differ significantly. This depends on environmental situation in various areas. External conditions and current infrastructure, where RES technologies could be applied, also differ; so spatial planning is the most appropriate tool for benefit assessment via urban energy development plans.

Most countries mainly support RES technologies on supply side, while opportunities to use solar energy are significant at consumers' side, such as small systems of solar collectors in multifamily houses, etc. Support for consumers is the most expedient via support programmes and not by supporting scattered consumers, as programmes enable achieving actual effect. Urban district renovation of buildings enables planned actions, which could solve technical and financial problems for district heating companies due to reduced consumption. The law on Renewable energy sources encloses several statements, dedicated for individual consumption of RES using energy generators. However, such schemes as feed-in tariffs, green certificates, auctions, etc., which are applied in Lithuania, are adopted for scale projects implementing companies with respective competencies. Applying of above schemes for small producers increase administration costs and is often not acceptable due to high complexity level. Here territorial administration is needed.

Periodization of economic age and assessment of various RES support forms, with regard to the specifics of each, has its own goals and tasks from macroeconomic attitude. We should distinguish short-term, medium-term and long-term goals and respective motivation systems for various RES technologies. Domination of few RES in power and heat generation is an evident shortcoming related to short-term perspective (which reflects new EU support period). Large imported wind plants are domineering in power generation, though small PV, wind and various hybrid plants will have wider application in medium and long-term perspective.

The main problem is that programmes for implementation of legally delegated functions have no consolidating unity. This causes serious problems, e.g., renovation of multifamily residential houses would reduce heat consumption significantly, and this would cause growth of district heating tariffs due to growing fixed heat generation and supply costs. Stable and balanced programme with carefully selected subsidizing, enabling interaction of consumers and heat suppliers, is necessary for solving such problems.

Various RES differ in various aspects; thus, classification based on renewability is important for assessment of possibilities and for justification of respective economic measures in terms of future implementation. The important shortage is that RES assessment and implementation takes place in short-term period, related to EU support period till 2020. This is the reason why only biomass and wind plants are dominating in heat and power generation. Thus, we should distinguish short-term (till 2020) and medium-term (till 2030) periods. Danish example shows that prospective vision (till 2050) is also possible. Infrastructure like district heating systems and spatial planning systems are very important for assessment of complex effect and practical implementation. Solar energy can be used for both power and heat generation. The scale of technologies is wide, and competitiveness dynamics is different bearing in mind the reduction of technological costs during the time. Use of solar collectors for production of hot tap water in multifamily residential houses is already competitive at present. In a long run, all RES technologies should be evaluated not only as competing with each other, but also as able to gradually replace fossil fuel. Fast sally of one or another technology is impossible: solar energy can be used during a limited summer period, and heat generation volumes are not large; the use of wind energy depends on climate conditions. Scattered generation character and the fact that most of such installations can be implemented on the demand side are left for private initiative. With regard to involving hardly competing energy sources, an investment managing institution is necessary. Municipalities could act as such institution seeking for development of wide scale RES technologies. Actual legislation in Lithuania shows that municipal powers and responsibilities regarding RES are regulated in detail. There are no barriers, except financing issues for municipalities' actions. Some of them have signed the Covenant of Mayors; some have adopted RES development action plans; however, only few of them are really acting.

Each RES has different value in terms of external benefit; thus, respective promotion schemes can and must differ significantly. This depends on environmental situation in a certain area. External conditions and current infrastructure for RES technologies differ; thus, spatial planning is the most appropriate tool for assessment of benefit, like urban energy development plans. All countries pay little attention to support of RES technologies on demand side. Support of consumers is most effective via programmes and not scattered consumers, as it provides possibilities to achieve an actual effect.

Technological, economic, social and environmental criteria are used for assessment of energy development. Technological criteria, the share is RES in primary energy consumption, heat and power generation and consumption, are among the most important. Economic criteria are related to the assessment of investment needs and value added. Necessary investment for technological installations, operation costs as well as planned income for product sales together with value added for the given region are to be assessed for energy development stages. Environmental indicators assess the index of sustainable process, which is based on the methodology of ecological footprint, which shows the area of global land area to cover energy needs of the consumers in the given region, and these needs are covered using respective technologies with estimated environmental impact.

Based on the methodology presented above, estimates are performed. The results of research are presented in section 5.4.2 below.

However, balancing of various RES implementation aspects is still important. Development stages are maximized based on technological indicators (RES share in power and heat sectors), economic indicators (change of created value added and investment volumes) and social indicators (employment or new jobs) and minimized based on environmental indicators (the least impact assessed using index of sustainable process).

Optimization of existing energy infrastructure forms short-term stage for energy sector development. Introduction of new energy technologies and smart grid control systems into the programme forms a long-term stage. Formation of stages can be performed via investment limitation. Such case of modelling can be used for planning the investment, using Lithuanian and EU funding for RES development. In case of unlimited investment in energy sector, one can focus on promotion of technological innovations in energy generation, using new sources in energy production and regarding ecologic, social and energy efficiency criteria.

Analysis of energy generation technologies, using RES, energy generation costs changes based on the type and origin of RES, on distance from extraction, collecting, processing and transportation of primary resources, on necessary investment in energy production, storage and transmission technologies. It was defined that carefully evaluated environmental, social and other hidden costs increase energy generation and fossil fuel costs. Increase of environmental costs could have positive impact on RES development, as it should stimulate consumers to choose RES using energy generation technologies and refuse to use fossil fuel. Growing fossil fuel prices due to introduction of hidden costs and social taxes would stimulate social economic development thus leading to faster development of RES actions, infrastructure and energy generation technologies (Duer, 2010).

Proper assessment of taxes and costs helps RES competition with fossil fuel; however, it should be mentioned that market stakeholders are not so much interested in the price, but rather in stability and possibility to forecast, which would provide background to invest in energy saving measures thus reducing total energy expenses. If saving of one kWh costs less than generating it, investment should be directed not towards installation of new capacities, but towards saving secondary energy thus providing energy services at the least costs. Investment should be directed to where it gives the highest effect, and investment in energy saving measures is often more beneficiary from economic as well as environmental point of view (Klevas, 2010).

5.4. Framework for Sustainable Energy Programmes

The methodological approach above is suitable for sustainability assessment using regional social-economic-environmental indicators, such as increase of security of energy supply, new jobs, new enterprises, value added, and greenhouse gas reduction. The methodology of integration sustainable energy projects into local/regional development programmes has been elaborated as a key to integration of sustainable energy projects into the regional deployment level (Klevas, Antinucci, 2004). Implementation of National sustainable energy strategy should be supported by sharing the tasks on the increased share of renewables and energy efficiency improvements between regions and authorities (Klevas, Minkstimas, 2004).

Local programmes could be the most suitable tool for redirecting urban areas towards sustainable energy. Municipalities have both, policy instruments and means, to implement programmes, and they are in direct contact with citizens and businesses. Sustainable energy policy linked to specific national measures could contribute to reducing and managing energy and environmental problems in general and greenhouse gases in particular. A very interesting example is the experience of the Netherlands. Municipal climate policy has been operating since 2001 in the Netherlands. Since then, more than 150 municipalities have made a start. Dutch municipal climate policy is based on a covenant between central government and representative bodies from the municipalities and provinces. The tasks are clearly defined: central government focuses on identifying the climate objectives, including basic standards, acts as a facilitator, while the municipalities will be the ones that actually do the work. The Ministry of Housing, Special Planning and the Environment (VROM) makes subsidies available so that municipalities can release an extra capacity for implementing the policy. The level of these subsidies is linked to the degree of ambition and the actual results achieved. A total of 37 million euro has been made available for the coming years.

Each municipality chooses its own issue: municipal buildings and installations, sustainable energy, etc. Before municipality selects working on one or more issues, it attaches a level of ambition to it. This serves to indicate how far the municipality wants to move on this issue. Distinctions are made between an active policy, a leading policy and an innovative policy. Based on the outcome, the municipality establishes its policy for the next five years. This policy is then implemented by means of a phased plan, which is also provided by the municipal climate policy programme.

Analysis of energy sector and the stages of its development are investigated using process network synthesis method with overview of the best practice examples in EU member-states.

Formation of energy sector development stages is based on the following statements:

1. Gradual reduction of the use of fossil fuel and respective increase of RES with regard to RES action plans in towns or regions.
2. Formation of energy sector development plants seeks to achieve major share of RES in energy production by 2020.
3. Formation of long-term energy sector development energy generation using RES and residual energy and transfer to power grids and district heating networks will be developed using smart grid technologies.
4. The following stages of energy sector development are investigated:

5. **Short-term** development stage encloses period till year 2020: this is RES programme, aiming to implement planned RES indicators for energy sector with rather moderate financing and technical tools. This is the stage of optimization of energy sector structure, including reformation of district heating sector, promotion of small-scale modernization of heat sector, replacing fossil fuel technologies by RES technologies. In fact, this is optimization of existing energy system by gradual modernization and wider use of feasible RES technologies.
6. **Long-term** energy sector development stage after 2020. This is the stage of optimal use of RES and smart grids, when existing energy system is reorganized via implementation of smart energy technologies enabling wider use of RES – solar, wind, geothermal with the aim to maximize RES energy supply.

Short-term period till 2020: this is the stage of energy sector development to be implemented till 2020 according to the national RES measures in the action plan, using national and EU financial support. This period envisages wider use of biomass for energy generation. This is development of biomass in DH sector by replacing existing natural gas boilers with biomass installations. Decentralised heat generation envisages modernization of thermal installations. Heat installations using low quality forest wood and wood waste are to be replaced by more efficient ones, using wood pellets, briquettes. The use of natural gas is unrestricted; however, the change of prices for natural gas and biomass follows, and using ITS software provides the feasibility estimate.

The promotion of RES on a wider scale was implemented in EU by creating green settlements or so-called eco villages. e.g., state funded project Climate Menu in the Netherlands enabled towns to select sectors, which needed exceptional attention for meeting national obligations. The share of subsidy assigned to certain municipality depends on its area, population; in this way, smaller municipalities co-operate and implement joint projects thus saving money. There are approximately 500 municipalities in the Netherlands, 300 of them have already consumed support. Each town-depending scale, organizational skills and specifics could promote private activity areas. State subsidies can cover 50% of project implementation costs, and the rest is covered by municipality. Every municipality selects financing sector. There are three levels of goals, which show the level at which municipality intends to develop the funded area. State funded projects should be implemented in four years. The use of RES is promoted in Austria in similar way. Every municipality selects individual support measures and financial tools promoting development of RES technologies. Mainly, these are solar technologies and modernization of district heating systems using biomass (Klevas, Bieksa and Murauskaite, 2014).

The global movement of green settlements appears to be very important. It was initiated by UN Habitat declaration in 1976. This idea had become rational and effective in various aspects, and energy use was one of the most significant. Following the principles of Agenda 21 accepted at Rio Summit in 1992, the baseline of sustainable development for settlements was adopted, and finally, the essential principles were adopted at Habitat-II UN conference in Istanbul in 1996.

Declaration on sustainable development of settlements states that cities and settlements were and are the incubators of the civilization and generators of growth, which have accelerated the evolution of knowledge, culture and traditions, as well industry and trade. Properly planned and controlled urban settlements create favourable perspectives for human development and protection of global nature resources.

The movement of eco villages or green settlements is driven by local initiatives, local communities and group leaders to pursue the main objectives: to create the better, healthier, eco-friendlier lifestyle, fostering the social and community lifestyle and encourage the development of ecological permaculture and nature economy (Birekland, 2008).

The main recommendation is to establish unified principles for the estimation of development of renewable energy sources technologies, which allow justifying the unified and long-term incentive system for the use of RES technologies based on economic theory, according to the following key aspects:

A developed method and calculation programme of the use of RES impact on macroeconomic indicators was the first *focus of the evaluation*. Moreover, grouping and classification of RES was done according to renewal nature: a) Renewal in the nature sphere means inexhaustibility (perpetual, never exhausted use) of natural RES, such as wind, solar, and geothermal energy. Furthermore, it means quality of inexhaustibility, which is one of the main accents of sustainable development of energy, ensuring a reliable supply of energy resources for the future generations. Wind, solar, geothermal energy in particular represents the natural phenomenon of renewal. This type of energy is not produced, it is perpetual; and thus, *the extent of the use of one period does not affect the extent of the use of the next period,* which affects the expectations of investors and consumers. Therefore, the important possibility of inexhaustibility of natural resources for the economic evaluation is the correct application of the procedure of discounting. The discount rate, which is a part of the price system, became a starting point for deliberations; b) Inexhaustibility is the product of more or less continuous economic activity. It is represented by the produced biomass and biofuel. This is a forestry and agricultural production. Great expectations currently are associated with the use of biomass for the production of electricity in competition with the wind, also the massive use for the production of heat. However, the result of economic activity and over-exploitation may respond to decline of national wealth of forest or farming lands; c) Inexhaustibility is a constant generation of waste in cities (municipal waste), in agriculture. The main effect of this type of RES is not as much energy as environmental.

The second *focus of the evaluation* was the motivation of RES technologies demand in short, medium, and long term. In macroeconomic terms, a periodization of economic period and evaluation of various forms of support for RES, as well as organization, taking into account the specifics of each period, has its own aim and objectives.

In the short term, which comprises the period before 2020, a motivational system was developed, legitimized as an assumption of achieving the objectives of the use of RES, as the EU directives oblige. However, the scheme of the public support should be more favourable in respect of solar, wind, and geothermal energy. Selection of assessment methods for the use of RES and links between macroeconomic indicators are developed, also an algorithm of calculations was formed. A formal mathematical assessment was carried out applying the formed calculation programme. The inventory of RES financial and economic forms of support was performed, emphasising their funding sources, result indicators, and dynamics of the achievement. Review of support schemes and measures of RES technologies financing, lending, pricing, etc. used in foreign countries was done. Moreover, funding opportunities of advanced energy technologies from international funds (such as JESSICA, JEREMIE, etc.) was evaluated in Lithuania. The origin of acceptable financing channels was evaluated, also the use of innovative financing instruments for the financing of end-users and industry. However, *in the medium term,* these measures must be adapted and adjusted by eliminating apparently distorting incentives of market relations. The role of Government should gradually be limited to the RES sector, considering the fact that environmental and social return of the investment in the renewable energy sector is significantly higher than the private (individual) return. The Government should form national programmes and support their implementation by providing subsidies, soft loans, loan guarantees, and tax incentives.

In the long term, the main emphasis should be given to the improvement of education system and scientific research. Impact of scientific and technical progress, as well as formation of various measures and assumptions do not immediately affect country's international competitiveness. Effects on intellectual capital and its growth are perceived in the near future. The most important form of financing should be *public-private partnership*, i.e., long-term cooperation based on agreement between institution of public sector and representative of private sector for the implementation of complex of activities (for example, infrastructure object's design, construction, reconstruction, repair and maintenance). The advantage of the partnership is that the risk of project implementation is distributed among the partners according to their abilities to manage risk.

The third aspect was evaluation of different motivation systems in accordance with energy suppliers, technology developers, and users. The research showed a significant disproportion between support for energy suppliers and energy consumers. Moreover, calculation proved that major macroeconomic effect is possible on the consumers' side.

On the other hand, there is a very important issue of balancing various forms of RES. The main approach is organized integration of various RES technologies within the programmes of cities energy, where the DH plays a key role. Currently, monopolistic position determines that DH is a particularly suspensive factor of the competitiveness for Lithuanian economy due to price volatility, due to steadily rising prices, due to reduction in the purchasing power of significant part of the population. The comparative analysis with the economic assumptions of DH systems in foreign countries led to the conclusion that DH drawbacks could be transformed into advantages by changing the pricing principles and by properly organising the state support system for RES.

CONCLUSION

1. Economic research of RES (renewable energy sources) intensified only in recent years through the implementation of European Union directives, with particular emphasis on the role of regions (cities, districts). Incentives for a wider use of RES are declared both in the European Union and in Lithuanian laws. Adoption of the Directive 2009/28/EC "on the promotion of the use of energy from renewable sources" required development of a National Renewable Energy Action Plan for 2010-2020 and other documents. This indicates that energy policy of the Republic of Lithuania gives increasingly important place to the development of RES. It can be said that there is a lack of research, which would answer questions about the new financing methods and economic assumptions that would link the economic policy on the macroeconomic level and would justify benefits of the use of RES technologies state-wide.
2. Renewable energy policy starts from a clear strategic idea. It is necessary to establish a link between strategic national and local goals and aims of the private business. Directives, legal acts, rules and regulation mechanisms for promotion of RES innovation and investments are necessary as a top-down tool to establish the best decisions on national scale. However, a bottom-up approach is necessary, because the best solutions usually come from local stakeholders and leaders. Balance is required seeking to avoid market distortions, which later arouse undesirable necessity for economic policy changes.
3. Statistical trends leave benefits of the individual project invisible to the country's macroeconomic indicators. They do not provide any tools or techniques enabling optimization of RES sector development

(or introducing support measures) for maximizing the macroeconomic benefits of choosing optimal support measures. Proposed microeconomic analysis of the project chosen by the individual investor (taking into account the economic benefits for the developer), utilizing the external parameters, affecting the macroeconomic indicators. Results obtained in a single project analysis can be interpolated to assess possible impact of the support scheme (country level).

4. It is necessary to have in mind that every country is in different situation of economic development. Therefore, the mechanical copy-paste approach of the economic policy is a negative phenomenon. RES support policy does not have to contradict the Lithuanian strategic aim, which is to reduce country's dependency upon the single energy supplier. In this respect, large-scale onshore and offshore WPP is a promising scenario for development of wind energy sector in the country.
5. The extent and directions of the use of energy resources can depend on decisions and liabilities of governmental programmes. The analysis of experience of developed countries disclosed that investment in the RES sector of these countries can come through government, banks, venture capital funds, private equity funds, pension funds, sovereign funds and other. RES policy measures and policy targets implemented to support RES sector contributes to increasing amount of private sector participation. Private financial resources come from debt market, equity market or capital market.
6. Results of the analysis revealed that the availability of financial channels depends on the stage of life cycle of RES sector development. Because of the high risk, technology research activity is mainly financed using public financial sources, such as grants or subsidies. When technologies move out of research to the development stage, the amount of public financing reduces. Venture capital and privative equity come into the place. Because of high level of risk and dismissal of the requirements held by financial institutions for receiving loans, at an early stage of technology development, RES technology developer is not capable of receiving financing from credit/debt markets. Financial sources of credit market are available when RES plants are constructed.
7. Each of RES types has a very broad and distinctive use scale, so each renewable has its own specific support. The investigation of this topic

focused on the potential analysis of wind, solar and geothermal energy in rather narrow applications only. Survey of energy classification, financial, economic assistance forms inventory, highlighting their sources of funding, performance indicators and the achievement of dynamics analysis was performed. It is recommended to do this for all RES types.

8. Several financing channels are developed, through which investment in the RES sector comes. The analysis revealed that government, multilateral development banks, regional development banks, bilateral agencies, commercial banks and EU are the most significant RES sector financing channels in developing countries. Due to the RES sector's ability to solve regional issues, and because of the established RES policy, participation of private sector increases. The results showed that the ability to use financial channels also depends on the income level, the type of RES, and technology implemented.
9. The traditional RES sector financing instruments used are subsidy, loan and equity. During the time of economic recession, the ability to receive loans reduced. Although Libor reduced, banks considered RES sector of high risk and were risk-averse; therefore, set high spreads for loan and reduced the loan repayment period. Nowadays, various modifications of traditional instruments are available. This is the case of mezzanine finance. Thus far, Green bonds and notes are issued. They all help to raise the capital and increase investment in RES.
10. Innovative financing instruments provided under JESSICA and JEREMIE initiatives as well investment subsidies are favourable for developing solar energy sector in Lithuania. Seeking to expedite solar sector development in Lithuania, it is essential to review feed-in tariff, which currently is too low and impedes implementation of solar PV technologies. Soft loan provided under the JESSICA initiative and investment subsidy given by EIF are instruments, which could contribute to the increased consumption of solar electricity in Lithuania. Policy makers should consider these instruments when preparing electricity sector development plans and organizing support schemes for development of solar PV technologies in Lithuania. Solar collectors for heat energy production are valid in small multifamily houses in Lithuania. It is rational to install solar collectors even under market conditions, using hard loans provided by banks. This shows that solar collectors could compete in the district heating market

without any support. However, if they are implemented at the end user point, they could create a higher (additional) district heating cost for the rest of the costumers.

11. Each of the types of RES has a very broad and distinctive range of applications; therefore, the attention is concentrated on the analysis of possible alternatives of RES. Research of wind energy classification, inventory of financial and economic forms of support, highlighting of their funding sources, performance indicators and the dynamics of their achievement was performed. An important assumption was the analysis of demand schemes of small wind power plants, serial production of which would be possible in Lithuania. The analysis of energy, economic, social, and environmental conditions was performed in the selected district (communities of Riese and Avizieniai). Moreover, evaluation of energy potential, also formation and estimation of development scenarios of energy sector was done.
12. One of the biggest technical challenges of wind energy utilization in comparison with traditional forms of electricity generation is that the output of WPPs directly depends on the volatility of the wind. For this reason, power system should maintain an adequate power reserve to compensate fluctuation of the wind speed. Therefore, additional burden of electricity balancing is put on the power system, especially if the wind power comprises a significant share of the country's power system.
13. LCOE calculations showed that under the current market conditions, a 2000 kW WPP could generate electricity below the market price. It was also estimated that neither new nor refurbished 250 kW WPP would be competitive in Lithuanian electricity market if no support scheme were applied. Small wind turbines with capacity below 10 kW would be economically attractive only with 30-50% support for investments.
14. The measures of medium-term RES promotional schemes should be adjusted and updated by refusing incentives that apparently distort relations in the market. In the area of wind electricity sector, the role of the Government should be limited to ensuring and establishing the infrastructure for supply of wind electricity (i.e., improvement and construction of electricity networks). The perspectives of other types of RES should be analysed in a similar way. An adoption of support schemes for promotion of demand of small-scale wind power plants, mass manufacturing of which would be possible in Lithuania, is of

high importance. The research results demonstrate that although presently small-scale WPPs are not economically competitive, these plants can be efficient in remote from the city areas. The hybrid wind-solar power plants could have a high demand in various territories.

15. It is acknowledged that it is easier to implement an integrated complex of environmental, social and economic policies in cities and districts. Public expenditure and funding from Structural Funds should, where possible, allow achieving more than one policy objective of sustainable development. Sustainable development of energy at the regional level requires support measures for organizing and implementing institutions. Municipalities can play an important role in promoting assumptions of sustainable energy development, because the local government performs a number of functions in the energy sector, and these opportunities should be used.
16. In the perspective, it is expected that green villages will develop faster, and their role will increase. One of the main objectives of such construct is self-supply of energy by using RES technologies. In the medium- and long-term perspectives green villages will become a venue for small-scale and hybrid RES technologies. The state choices of the strategic energy development directions will influence the possibilities to construct large-scale on-shore and offshore WPPs.

REFERENCES

Acker, T. L.; Williams, S. K.; Duque, P. N.; Brummels, G. & Buechler, J. 2007. Wind resource assessment in the state of Arizona: Inventory, capacity factor, and cost. *Renewable Energy*. Volume 32, Issue 9, p. 1453-1466.

Act on the Promotion of Renewable Energies in the Heat Sector, 2008 // http://www.bmu.de/files/english/pdf/application/pdf/ee_waermeg_en.pdf

Act on the Promotion of Renewable Energies in the Heat Sector: Consolidated, non-binding version of the Act with amendments arising from the "Act Implementing European Renewable Energies Legislation," 2011 // http://www.bmu.de/files/english/pdf/application/pdf/ee_waerme_2010_en.pdf

Al-mulali, U.; Fereidouni, H.; Lee, Y.; Sab, Ch., 2013. Examining the bi-directional long run relationship between renewable energy consumption and GDP growth. *Renewable and Sustainable Energy Reviews*, Volume 22, p. 209-222.

Analysis of wind power development possibilities, 2009. Report of the 1st stage. Kaunas University of Technology, Lithuanian Energy Institute. p. 149.

Antinucci, M., De Borchgrave, C., Henaux, C., & Klevas, V. 2002. EU Support for Formation of Regional Energy Development: Problems and Possible Solutions. P*aper presented at the 2nd International conference, Klaipeda*, Lithuania.

Apergis, N. & Payne, J. E. 2010a. Renewable energy consumption and economic growth: Evidence from a panel of OECD countries. *Energy Policy*. Volume 38, Issue 1, p. 656-660.

Apergis, N. & Payne, J. E. 2010b. Renewable energy consumption and growth in Eurasia. *Energy Economics*. Volume 32, Issue 6, p. 1392-1397.

Apergis, N. & Payne, J. E. 2011. The renewable energy consumption–growth nexus in Central America. *Applied Energy*. Volume 88, Issue 1, p. 343-347.

Apergis, N. & Payne, J. E. 2012. Renewable and non-renewable energy consumption-growth nexus: Evidence from a panel error correction model. *Energy Economics*. Volume 34, Issue 3, p. 733-738.

Asian Development Bank, 2012. Clean Energy Program: Accelerating Low-Carbon Development in Asia and the Pacific Region // accessed 15.06.2015 http://www.adb.org/sites/default/files/pub/2012/clean-energy-program-brochure.pdf.

Augutis J, Krikstolaitis R, Peciulyte S, Konstantinaviciute I. Sustainable development and energy security level after Ignalina NPP shutdown. *Technological & Economic Development of Economy*. 2011; 17:5-21.

Bassi, A.M., Yudken, J.S., Ruth, M., 2009. Climate policy impacts on the competitiveness of energy intensive manufacturing sectors. *Energy Policy* 37 (8), 3052-3060.

Beckerman W. "Sustainable development": is it a useful concept? // *Environmental Values*. 1994. N3. p. 191-209.

Behrens, A., 2009. The Financing of the Global Energy Efficiency and Renewable Energy Fund (GEEREF). Centre for European Policy Studies // accessed 10.07.2015 http://www.undpcc.org/undpcc/files/docs/publications/1862.pdf.

Benkovic, S., Makojevic, N., Jednak, S., 2013. Possibilities for development of the Electric Power Industry of Serbia through private source financing of small hydropower plants. *Renewable Energy*. Volume 50, p. 1053-1059.

Bildirici, M. 2013. Economic growth and biomass energy. *Biomass and Bioenergy*. Volume 50, p. 19-24.

Birekland, J. 2008. *Positive development from vicious circles to virtuous cycles through built environment design*. London: Sterling VA.

Blazejczak J., Braun F. G., Edler D., Schill W.P., 2011 Economic Effects of Renewable Energy Expansion A Model-Based Analysis for Germany, Deutsches Institut für Wirtschaftsforschung Discussion Papers 1156.

Bloomberg New Energy Finance, 2012. Global Trends in Renewable Energy Investment 2012–Data Pack. Frankfurt School of Finance & Management.

Blum, N. U.; Wakeling, R. S.; Schmidt, T. S. 2013. Rural electrification through village grids–Assessing the cost competitiveness of isolated renewable energy technologies in Indonesia. *Renewable and Sustainable Energy Reviews.* Volume 22, p. 482-496.

Bobinaite V., Juozapaviciene A., Konstantinaviciute I. 2011. Assessment of Causality Relationship between Renewable Energy Consumption and Economic Growth in Lithuania. *Inzinerine Ekonomika-Engineering Economics,* 2011, 22(5), 510-518

Bobinaite, V. & Tarvydas, D. 2014. Financing instruments and channels for the increasing production and consumption of renewable energy: Lithuanian case. *Renewable and Sustainable Energy Reviews*, Vol. 38, p. 259-276.

Branker, K.; Pathak, M. J. M. & Pearce, J. M. 2011. A Review of Solar Photovoltaic Levelized Cost of Energy. *Renewable & Sustainable Energy Reviews*, 15, p. 4470-4482.

Brown M. 2001. Market Failures and Barriers as a Basis for Clean Energy Policies. // *Energy Policy* 29 (14): 1197-1207.

Bruggink, J., 2012. Energy aid in times of climate change: Designing climate compatible development strategies // accessed 12.05.2015 http://www.ecn.nl/docs/library/report/2012/e12006.pdf.

Bürer, M. J., Wüstenhagen, R., 2009. Which renewable energy policy is a venture capitalist's best friend? Empirical evidence from a survey of international clean-tech investors. *Energy Policy*, Volume 37, Issue 12, p. 4997-5006.

Canadian Renewable Energy Alliance, 2006. Financing Sources and Mechanisms for Renewable Energy and Energy Efficiency // accessed 08.11.2014 http://www.canrea.ca/pdf/CanREAFinancingPaper.pdf.

Cansino JM, Pablo-Romero MdP, Roman R, Yniguez R. Promoting renewable energy sources for heating and cooling in EU-27 countries. *Energy Policy*. 2011; 39:3803-12.

Chien, T., Hu, J. L., 2007. Renewable energy and macroeconomic efficiency of OECD and non-OECD economies. *Energy Policy* 35, 3606-3615.

Chien, T., Hu, J. L., 2008. Renewable energy: An efficient mechanism to improve GDP. *Energy Policy* 36, 3045-3052.

Ciarreta, A.; Gutiérrez-Hita, C & Nasirov, Sh. 2011. Renewable energy sources in the Spanish electricity market: Instruments and effects. *Renewable and Sustainable Energy Reviews*, Vol. (15), p. 2510–2519.

Ciegis, R. *Natural resources and environmental economics*. 2009, Klaipeda, p. 771.

Ciegis, R., & Streimikiene, D. 2005. Integration of sustainable development indicators in development programmes. *Engineering economics*, N. 2, pp. 28-41.

Croce, C., Kaminker, R., Stewart, F., 2011. The Role of Pension Funds in Financing Green Growth Initiatives, OECD Publishing, Paris. // accessed 08.07.2015 http://climatebonds.net/wp-content/uploads/2011/09/OECD_Role_of_PFs_in_Financing_GreenGrowth-WP10.pdf

Das Bundesamt für Wirtschaft und Ausfuhrkontrolle (BAFA, 2012) // http://www.bafa.de/bafa/de/energie/erneuerbare_energien/index.html

De Moor, A., 2001. Towards a grand deal on subsidies and climate change. *Natural Resources Forum* 25, p. 167-176.

Domac, J., Richards, K., Risovic, S., 2005. Socio-economic drivers in implementing bioenergy projects. *Biomass and Bioenergy* 28, 97-106.

Duer, H., Christensen, P.O. Socio-economic aspects of different biofuel development pathways. *Biomass and bioenergy*, 2010, 34, p. 237-243.

Dusonchet, L., Telaretti, E., 2010a. Economic analysis of different supporting policies for the production of electrical energy by solar photovoltaics in eastern European Union countries. *Energy Policy*. Volume 38, Issue 8, p. 4011-4020.

Dusonchet, L., Telaretti, E., 2010b. Economic analysis of different supporting policies for the production of electrical energy by solar photovoltaics in western European Union countries. *Energy Policy*. Volume 38, Issue 7, p. 3297-3308.

EC, 2013. European Commission guidance for the design of renewable energy support schemes//Commission staff working document.

Ecofys, Fraunhofer ISI, Energy Economics Group, Lithuanian Energy Institute, 2011. *Renewable Energy Policy*: *Country profiles*. // http://www.reshaping-res-policy.eu/downloads/RE-Shaping_CP_final_18JAN2012.pdf

Environmental Investment Fund, 2013 // accessed 02.05.2015 http://www.laaif.lt.

Environmental Protection Agency of USA, 2009. American Recovery and Reinvestment Act of 2009: A Guide to Renewable Energy and Energy Efficiency Opportunities for Local and Tribal Governments // accessed 08.08.2015 http://www.epa.gov/statelocalclimate/documents/pdf/local_guide_to_arra.pdf.

ESTIF, Solar Thermal markets in Europe 2011, trends and Market Statistics 2011, http://www.estif.org/fileadmin/estif/content/market_data/

downloads/Solar%20Thermal%20Markets%20in%20Europe%20-%20 Trends%20and%20Market%20Stat.pdf

EU structural funds // accessed 02.08.2015 www.esparama.lt.

EURELECTRIC. Closing the circle of competitiveness: the need to reorient European electricity policy, 2004.

EUROFIDEME, 2013. Investment Fund dedicated to renewable energy project finance in Europe // accessed 10.07.2015 http://www.eurofideme2.com/ en_ eurofideme-2_investment-fund_33_39.html

European Commission, 2014. State aid: Commission adopts new rules on public support for environmental protection and energy // http:// europa.eu/rapid/press-release_IP-14-400_en.htm.

European Commission [EC]. ExternE: Externalities of Energy. Luxembourg: Office for Official Publications of the European Communities, 1995: pp. 11-15.

European Commission, 2012. Report from the Commission to the European Parliament and the Council On the Implementation of the European Energy Programme for Recovery // accessed 16.02.2015 http://ec.e uropa.eu/energy/eepr/doc/com_2012_0445_en.pdf.

European Commission. Directive 2009/28/EC of the European Parliament and of the Council of 23 April 2009 on the promotion of the use of energy from renewable sources and amending and subsequently repealing Directives 2001/77/EC and 2003/30/EC. 2009. *Official Journal of the European Communities*, 5.6.2009, L 140/16.

European Commission. Green book about energy efficiency. Luxembourg, 2005.

European Commission's report. 2004. http://europa.eu.int/eur- lex/pri/en/ dpi/cnc/doc/2004/com2004_0029en02.doc

European Parliament, 2009. Regulation (EC) No 663/2009 of the European Parliament and of the Council of 13 July 2009 establishing a programme to aid economic recovery by granting Community financial assistance to projects in the field of energy // accessed 16.07.2015 http://eur-lex.europa.eu/LexUriServ/LexUriServ.do?uri=OJ:L:2009:200:0031:0045: EN:PDF.

European wind energy association (EWEA). *The economics of wind energy*. 2009.

ExternE: Externalities of energy // European Commission, Directorate General XII, Science Research and Development, Luxembourg, 1995, pp. 266-36.

Fang, Y. 2011. Economic welfare impacts from renewable energy consumption: The China experience. *Renewable and Sustainable Energy Reviews.* Volume 15, Issue 9, p. 5120-5128.

Feasibility study on the integration of renewable energy power plants to the 330-110 kV transmission grid till 2030, 2014. Kaunas University of Technology, SC Litgrid, Vilnius.

Federal Ministry for the Environment, Nature Conservation and Nuclear Safety, 2009. *The Global Financial Crisis and Its Impact on Renewable Energy Finance.*

Ferderer J. P. 1996. Oil Price Volatility and the Macroeconomy: A Solution to the Asymmetry Puzzle.// *Journal of Macroeconomics* 18, 1-16.

Figge F., Hahn T. Sustainable value added-measuring corporate contributions to sustainability beyond eco-efficiency. *Ecological Economics* 2004; 48: 173-187.

Fournier, G., Rasmussen, D., 1986. Targeted Capital Subsidies and Economic Welfare. Cato Journal, Vol. 6, No, 1, p. 295-315 // accessed 11.07.2015 http://www.cato.org/sites/cato.org/files/serials/files/cato-journal/1986/5/cj6n1-14.pdf.

Frankfurt School of Finance & Management, 2012. Global Trends in Renewable Energy Investment 2012. Frankfurt School of Finance & Management.

Gass, V.; Schmidt, J.; Strauss, F. & Schmid, E. 2013. Assessing the economic wind power potential in Austria. *Energy Policy*, Volume 53, p. 323-330.

Genutis A. The effect of technological air heating by a solar collector without a transparency cover.//Proceedings of the 9th international Conference "Ecological energy resources in agriculture," 2004. p. 54-60.

Global Energy Efficiency and Renewable Energy Fund // accessed 15.06.2015 www.geeref.com.

Global Village Energy Partnership//accessed 17.08.2015 www. gvepinternational.org.

Gouldson A., Peter R. Integrating Environment and Economy. Strategies for local and regional government. London and New York: Routledge, 2000.

Griffin J. M., Steel H. B. *Energy Economics and Policy*. Orlando: Academic Press, 1986.

Grifth-Jones, S., Ocampo, J. A., Spratt, S., 2011. *Financing Renewable Energy in Developing Countries: Mechanisms and Responsibilities.*

Hanley N., Mourato S., Wright R. E. Choice Modelling Approaches: A Superior Alternative for Environmental Valuation? *Journal of Economic Surveys* 2001; 15(3): 435-462.

Heijungs R., Guinee J. B. Environmental life cycle assessment of products. Leiden: Centrum of Milieukund, 1992.

Hernández-Moro, J. & Martínez-Duart, J. M. 2013. Analytical model for solar PV and CSP electricity costs: Present LCOE values and their future evolution. *Renewable and Sustainable Energy Reviews*. Vol. 20, p. 119-132.

Hofman, D., Huisman, R., 2012. Did the financial crisis lead to changes in private equity investor preferences regarding renewable energy and climate policies? *Energy Policy*. Volume 47, p. 111-116.

Hotteling H. 1931. The economics of exhaustible resources // Journal of political Economy, N. 39. p. 137-175.

Implementation Plan for National Strategy on Development of Renewable Energy Sources during years 2010-2015, 2010 // http://www3.lrs.lt/pls/inter3/dokpaieska.showdoc_l?p_id=377162.

Inter–American Development Bank, 2011. Integrated Strategy for Climate Change Adaptation and Mitigation, and Sustainable and Renewable Energy // accessed 15.08.2015 http://idbdocs.iadb.org/wsdocs/getdocument. aspx?docnum=35802849.

Inter–American Development Bank, 2013. Latin America and the Caribbean could cover all their electricity needs using renewable resources: IDB report // accessed 15.05.2015 http://www.iadb.org/en/news/webstories/2013-06-18/renewable-energy-in-latin-america-and-the-caribbean,10486.html.

International Energy Agency (IEA), 2010. Energy technology perspectives 2010: scenarios and strategies to 2050. Paris, France: International Energy Agency, IEA/OECD; 2010. p. 1-706.

IRENA, 2012. Power sector. Volume 1. // http://www.irena.org/DocumentDownloads/Publications/RE_Technologies_Cost_Analysis-WIND_POWER.pdf.

Jalilvand J., 2012. Renewable Energy for the Middle East and North Africa Policies for a Successful Transition // accessed 16.11.2014 http://library.fes.de/pdf-files/iez/08959.pdf.

Jaramillo, O. A.; Saldaña, R. & Miranda, U. 2004. Wind power potential of Baja California Sur, México. *Renewable Energy*. Volume 29, Issue 13, p. 2087-2100.

Justice, S., 2009. Private financing of Renewable Energy – A Guide for Policymakers // accessed 15.11.2014 http://sefi.unep.org/fileadmin/media/sefi/docs/publications/ Finance_guide_FINAL-.pdf.

Kalamova, M.; Kaminker, C., Johnstone N., 2011. Sources of Finance, Investment Policies and Plant Entry in the Renewable Energy Sector. OECD Environment Working Papers, No. 37, OECD Publishing. // accessed 06.11.2014 http://www.oecd-ilibrary.org/docserver/download/5kg706801lhb.pdf?expires=1360660915&id=id&accname=guest&checksum=372889DFF45BC66873C4D4BE29175607

Kaminker, Ch., Stewart, F., 2012. The Role of Institutional Investors in Financing Clean Energy. OECD Working Papers on Finance, Insurance and Private Pensions, No. 23, OECD Publishing // accessed 06.04.2015 http://www.oecd.org/environment/WP_23_TheRoleOfInstitutionalInvestorsInFinancingCleanEnergy.pdf.

Keyuraphan, S., Thanarak, P., Ketjoy, W., Rakwichian, W., 2012. Subsidy schemes of renewable energy policy for electricity generation in Thailand. Procedia Engineering. Volume 32, 2012, Pages 440-448.

Klevas V., & Antinucci M. 2004. Integration of national and regional energy development programs in Baltic States. *Energy Policy*, Vol. 32, pp. 345-355.

Klevas V., Bieksa K., Kleviene A., & Stankevicius M. 2010. Regional approach of sustainable energy development // 11th IAEE European conference on energy economy, policies and supply security: surviving the global economic crisis, Vilnius, Lithuania, August 25-28, International Association for Energy Economics, Cleveland OH, 2010. ISSN 1559-792X, pp. 1-10.

Klevas V., Minkstimas R. 2004. The guidelines for state policy of energy efficiency in Lithuania. *Energy Policy* 2004; vol. 32: 309-320.

Klevas V., Streimikiene D. 2006. Lietuvos energetikos ekonomikos pagrindai. Kaunas, LEI, 2006, p. 404.

Klevas V., Streimikiene D., & Kleviene A. 2009. Sustainability assessment of the energy project implementation in regional scale. *Renewable & Sustainable Energy Reviews*. ISSN 1364-0321.- vol. 13.- N.1, pp. 155-166.

Klevas V., Streimikiene D., Skalisius V. 2006. Sustainability assessment of the energy projects integration in regional scale // Securing energy in insecure times. Proceedings of 29th IAEE International Conference, Potsdam, Germany, 7-9 June 2006, Cleveland OH: International association for Energy Economics, 2006, ISSN1559-792X, p. 69-70.

Klevas, V., & Antinucci, M. 2004. Integration of national and regional energy development programs in Baltic States. *Energy Policy*, Vol. 32, pp. 345-355.

Klevas, V., 2010. Regional approach for policies and measures aiming to sustainable energy development. Paths to sustainable energy. Croatia: InTech, 117-132. ISBN 978-953-307-401-6.

Klevas, V., Bieksa, K., Kleviene, A., Bubeliene, J., Stankevicius, M., 2010. Principles of energy development sustainability evaluation. *Journal of Power Engineering,* 56, 92-102.

Klevas, V., Bieksa, K., Murauskaite, L., 2014. Innovative method of RES integration into the regional energy development scenarios. *Energy Policy* 64, 324–336.

Knapek, J. 2010. Effectiveness of feed-in tariff scheme – Lessons learned from Czech Republic // 11th IAEE European conference on energy economy, policies and supply security: surviving the global economic crisis, Vilnius, Lithuania, August 25-28. International Association for Energy Economics, Cleveland OH, 2010. ISSN 1559-792X, pp. 1-10.

KPMG, 2011. *Taxes and Incentives for Renewable Energy* // accessed 11.11.2014 http://www.kpmg.com/Global/en/IssuesAndInsights/ArticlesPublications/Documents/*Taxes-Incentives-Renewable-Energy*-2011.pdf

Krautkraemer J. A. 1988. The rate of discount and the preservation of natural environments // natural resource Modeling. No. 2, p. 421-437.

Kveselis V., Dzenajaviciene E. F., Masaitis S., 2011. The role of district heating and cooling technologies in energy provisions for building sector: challenges and perspectives. *Journal of Environmental Engineering*, 2, 762-768.

Laes E, Verbruggen A. Decision support for national sustainable energy strategies in an integrated sustainability assessment framework. Paths to sustainable energy. Croatia: InTech; 2010. p. 93-118.

Lantz, E., Doris, E., 2009. State Clean Energy Practices: Renewable Energy Rebates // accessed 11.11.2014 http://www.nrel.gov/docs/fy09osti/45039.pdf.

Law on Renewable Energy Sources, 2011 // http://www3.lrs.lt/pls/inter3/dokpaieska.showdoc_l?p_id=398874.

Lee, D., Orr, D. 1975. The private discount rate and resources "conservation" // *Can. J. Econ.* 1975, N8, p. 351-356.

Levitt, A. C.; Kempton, W.; Smith, A. P.; Musial, W. & Firestone, J. 2011. Pricing offshore wind power. *Energy Policy*. Vol. 39, p. 6408-8421.

Liming, H., 2009. Financing Rural Renewable Energy: A Comparison between China and India. *Renewable and Sustainable Energy Reviews*. Volume 13, Issue 5, p. 1096-1103.

Lithuanian Energy Consultants Association, 2011. *Causality analysis of* district heating *prices in Lithuanian municipalities*. Vilnius, 2011. [in Lithuanian].

Lithuanian National Renewable Energy Action Plan, 2009 // http://ec.europa.eu/energy/renewables/action_plan_en.htm

Lithuanian Statistics, 2013 ////accessed 18.07.2015 http://www.osp.stat.gov.lt/web/guest/statistiniu-rodikliu-analize1.

Louviere J. J., Hensher D. A., Swait J. D. *Stated choice Methods; Analysis and applications,* Cambridge. UK: Cambridge University Press, 2000.

LR Ministry of Agriculture. Rules of funding for the production of biofuel // http://www3.lrs.lt/pls/inter3/dokpaieska.showdoc_l?p_id=325586&p_query=&p_tr2=

Lu, X.; Tchou, J.; McElroy, M. B. & Nielsen, Ch. P. 2011. The impact of Production Tax Credits on the profitable production of electricity from wind in the U.S. *Energy Policy.* Volume 39, Issue 7, p. 4207-4214.

Lukosevicius, V., 2013. Lithuanian district heating sector in the light of EU directives. *Conference of Thermal Energy and Technology.* Lithuania, Kaunas, 2013, 31 January-1 February. [in Lithuanian].

Magnani, N.; Vaona, A. 2013. Regional spillover effects of renewable energy generation in Italy. *Energy Policy.* Volume 56, p. 663-671.

Mainali, B., Silveira, S., 2011. Financing off-grid rural electrification: Country case Nepal. *Energy.* Volume 36, Issue 4, p. 2194-2201.

Markandya A., Longo A. Identification of Options and policy Instruments for the Internalization of External Costs of electricity Generation. Social Science *Research Network Electronic Paper Collection* 2005: http://ssrn.com/abstract=729704.

Masdar Capital, 2013 // accessed 06.11.2014 http://www.masdar.ae/en/investment/detail/masdar-capital1.

Menegaki, A. N. 2011. Growth and renewable energy in Europe: A random effect model with evidence for neutrality hypothesis. Energy Economics. Volume 33, Issue 2, p. 257-263.

Mergers inquisitions, 2013 // accessed 16.04.2015 http://www.mergersandinquisitions.com/private-equity-vs-venture-capital/.

Michael J., Novogradac, C., 2011. Tax Credits are More Efficient Than Cash Grants. *Journal of Tax credits*, November 2011, Volume II, Issue XI // accessed 11.11.2014 http://www.novoco.com/journal/2011/11/news_ww_201111.php.

Miglo, A., 2010. Project Financing Versus Corporate Financing Under Asymmetric Information. *Journal of Business & Economics Research*. 8 (8). p. 27-42.

Ministry of Economy of the Republic of Lithuania. *National energy strategy*, Vilnius. 2002.

Ministry of Energy of the Republic of Lithuania, 2011. *Annual report of Ministry of Energy of the Republic of Lithuania*. Vilnius, 2011. [in Lithuanian].

Ministry of the Environment of the Republic of Lithuania. National Sustainable development strategy, Vilnius, 2002.

National Commission for Energy Control and Prices, 2011. 2011 July 29 resolution No. O3-233 Methodology of electricity from RES tariffs.

National Control Commission for Energy and Prices, 2013. Approved feed-in tariffs // accessed 02.06.2015 http://www.regula.lt/lt/atsinaujinantys_istekliai/tarifai/.

Norgard J. S. 2001. Sustainable energy future – Nordic perspective. Keynote at Pan-European NGO Sustainable Energy Seminar, Denmark, 2001.

Olmos, L., Ruester, S., Liong, S., 2012. On the selection of financing instruments to push the development of new technologies: Application to clean energy technologies. Energy Policy. Volume 43, p. 252-266.

Palojarvi, A. P. 2013. *Experiences with ecological technologies and practices.* (A. P. Palojarvi, Ed.) Vilnius, Lithuania: BMK.

Pearce D. W., Markandya A., Barbier E. B., 1989. Blueprint for a green economy, London: Earthscan.

Perednis E. Investigation of potentialities of the passive solar heating systems // *Proceed. of the Third ISES Europe Solar Congress Eurosun* 2000, 19-22 June 2000. – Copenhagen, Denmark, 2000. p. 161-165.

Perednis E. The system of buildings direct heating by solar energy and its feasibility study// Renewable energy in agriculture: proceedings of the international conference, Raudondvaris, Lithuania, September 16-17, 1999a. Raudondvaris: Institute of Agricultural Engineering LUA, 1999. ISBN 9986-732-13-1, p. 123-130.

Perednis E., Direct solar heating for buildings // *Energetika*. ISSN 0235-7208. 1999b. Nr. 4, p. 56-60.

Ramadhan, M. & Naseeb, A. 2011. The cost benefit analysis of implementing photovoltaic solar system in the state of Kuwait. *Renewable Energy*. Volume 36, Issue 4, p. 1272-1276.

Reichelstein, S. & Yorston, M. 2013. The prospects for cost competitive solar PV power. *Energy Policy*. Volume 55, p. 117-127.

REN21 MAP, 2011. Renewables Interactive Map: Renewables Policy Table, *Renewable Energy Promotion Policies*.

Renewable Energy and Energy Efficiency Partnership (REEEP), 2012. REEEP Strategy Overview 2012 to 2015 // accessed 15.05.2015 http://www.reeep.org/sites/default/files/REEEP%20Strategy%20Overview%202012-2015.pdf.

Rezessy, S., Bertoldi, P., 2010. Financing Energy Efficiency: Forging the Link between Financing and Project Implementation. Report prepared by the Joint Research Centre of the European Commission // accessed 15.08.2015 http://ec.europa.eu/energy/efficiency/doc/financing_energy_efficiency.pdf

RUSE Newsletter, 2005. No. 1. www.ruse_europe.org.

Sadorsky, P. 2009a. Renewable energy consumption and income in emerging economies. *Energy Policy*, Volume 37, Issue 10, p. 4021-4028.

Sadorsky, P. 2009b. Renewable energy consumption, CO_2 emissions and oil prices in the G7 countries. *Energy Economics*. Vol. 31, p. 456-462.

Scholer, K., Tan, M. M. L., 2012. Financing renewable energy projects in the US: case studies and opportunities // accessed 15.05.2015 http://www.lexology.com/library/detail.aspx?g=4f6a4099-bda6-4042-a1be-fae4f6011ca1.

Solar district heating platform // http://www.solar-district-heating.eu

Stankevicius, V., The fairest mean for renovation – the incentive. Report, LSTA, 2012. p. 36.

Statistics Lithuania, 2011. *Energy consumption in households* 2009. Vilnius.

Statistics Lithuania, 2015. *Energy consumption in households* 2014. Vilnius.

Strategy for Development of Renewable Energy, 2010 // http://www3.lrs.lt/pls/inter3/dokpaieska.showdoc_l?p_id=376097.

Swift, K. D. 2013. A comparison of the cost and financial returns for solar photovoltaic systems installed by businesses in different locations across the United States. Renewable Energy. Volume 57, p. 137-143.

Tahvonen O., Kuuluvainen J. 1993. Economic growth, pollution and renewable resources // *J. Environ. Econ. Manage*. Nr.24, P. 101-118.

Ting-Huan Chang; Chien-Ming Huang & Ming-Chih Lee, 2009. Threshold effect of the economic growth rate on the renewable energy development from a change in energy price: Evidence from OECD countries. *Energy Policy*. Volume 37, Issue 12, p. 5796-5802.

Treivel R., Partidario M. *The Practice of* Strategic Environmental Assessment, London: Earthscan, 1996.

Tugcu, C.; Ozturk, I.; Aslan, A. 2012. Renewable and non-renewable energy consumption and economic growth relationship revisited: Evidence from G7 countries. *Energy Economics.* Volume 34, Issue 6, p. 1942-1950.

United Nations Environment Programme (UNEP), 2005. Public Finance Mechanisms to Catalyze Sustainable Energy Sector Growth.

United Nations Environmental Programme (UNEP), 2013. Rural Energy Enterprise Development (REED) // accessed 15.08.2015 http://www.unep.org/climatechange/finance/SeedCapital/REED/tabid/29554/Default.aspx.

Van Ireland Ekko C., van der Straaten J., Vollebergh H. R. J. *Economic Growth and Valuation of the Environment.* Cheltenham, UK, Northampton, MA, USA: Edward Elgar, 2001.

Vos, R., Sawin, J., 2012. Chapter Ten – Policies for Financing Renewables. Renewable Energy Action on Deployment, p. 159-175.

Wang, X.; Kurdgelashvili, L.; Byrne, J. & Barnett, A. 2011. The value of module efficiency in lowering the levelized cost of energy of photovoltaic systems. *Renewable and Sustainable Energy Reviews.* Volume 15, Issue 9, p. 4248-4254.

World Bank, 2013a. Affiliates //accessed 15.08.2015 http://web.worldbank.org/WBSITE/EXTERNAL/EXTABOUTUS/0,,contentMDK:20040612~menuPK:8336267~pagePK:51123644~piPK:329829~theSitePK:29708,00.html.

World Bank, 2013b // accessed 15.08.2015 http://web.worldbank.org/WBSITE/EXTERNAL/TOPICS/EXTENERGY2/0,,contentMDK:21651596~menuPK:4140787~pagePK:210058~piPK:210062~theSitePK:4114200,00.html.

World Bank, 2013c. Financing renewable energy Options for Developing Financing Instruments Using Public Funds // accessed 15.08.2015 https://www.climateinvestmentfunds.org/cif/sites/climateinvestmentfunds.org/files/SREP_financing_instruments_sk_clean2_FINAL_FOR_PRINTING.pdf.

Wüstenhagen, R., Menichetti, E., 2012. Strategic choices for renewable energy investment: Conceptual framework and opportunities for further research. Energy Policy. Volume 40, p. 1-10.

Yildirim, E.; Saraç, S.; Aslan, A. 2012. Energy consumption and economic growth in the USA: Evidence from renewable energy. Renewable and Sustainable Energy Reviews. Volume 16, Issue 9, p. 6770-6774.

INDEX

F

G

H

I

J

L

M

T

V

W